T0213581

Processing-in-Memory for AI

Joo-Young Kim • Bongjin Kim
Tony Tae-Hyoung Kim

Editors

Processing-in-Memory for AI

From Circuits to Systems

 Springer

Editors
Joo-Young Kim
Korea Advanced Institute of Science and
Technology
Daejeon, Korea (Republic of)

Bongjin Kim
University of California
Santa Barbara, CA, USA

Tony Tae-Hyoung Kim
Nanyang Technological University
Nanyang, Singapore

ISBN 978-3-030-98780-0   ISBN 978-3-030-98781-7   (eBook)
https://doi.org/10.1007/978-3-030-98781-7

This Springer imprint is published by the registered company Springer Nature Switzerland AG
The registered company address is: Gewerbestrasse 11, 6330 Cham, Switzerland

Contents

Chapter 1
Introduction

Joo-Young Kim

1.1 Hardware Acceleration for Artificial Intelligence and Machine Learning

Artificial intelligence (AI) and machine learning (ML) technology enable computers to mimic the cognitive tasks believed to be what only humans can do, such as recognition, understanding, and reasoning [1]. A deep ML model named AlexNet [2], which uses eight layers in total, won the famous large-scale image recognition competition called ImageNet by a significant margin over shallow ML models in 2012. Since then, deep learning (DL) revolution has been ignited and spread to many other domains such as speech recognition [3], natural language processing [4], virtual assistance [5], autonomous vehicle [6], and robotics [7]. With significant successes in various domains, DL revolutionizes a wide range of industry sectors such as information technology, mobile communication, automotive, and manufacturing [8]. However, as more industries adopt the new technology and more people use it daily, we face an ever-increasing demand for a new type of hardware for the workloads. Conventional hardware platforms such as CPU and GPU are not suitable for the new workloads. CPUs cannot cope with the tremendous amount of data transfers and computations required in the ML workloads, while GPUs consume large amounts of power with high operating costs.

AI chip or accelerator is the hardware that enables faster and more energy-efficient processing for AI workloads (Fig. 1.1). Over the past few years, many AI accelerators have been developed to serve the new workloads, targeting from battery-powered edge devices [9–11] to datacenter servers [12]. As McKinsey predicted in the report [13], the AI semiconductor industry is expected to grow 18–19%

J.-Y. Kim (✉)
School of Electrical Engineering (E3-2), KAIST, Daejeon, South Korea
e-mail: jooyoung1203@kaist.ac.kr

© The Author(s), under exclusive license to Springer Nature Switzerland AG 2023
J.-Y. Kim et al. (eds.), *Processing-in-Memory for AI*,
https://doi.org/10.1007/978-3-030-98781-7_1

1

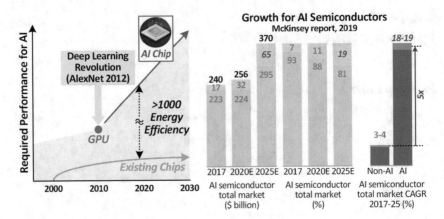

Fig. 1.1 AI chip and its market prediction

every year to 65 billion, accounting for about 19% in the entire semiconductor market in 2025. So far, the AI hardware industry is led by big tech companies. Google developed their own AI chip named tensor processing unit (TPU) that can work with TensorFlow [14] software framework. Amazon developed Inferentia chip [15] for high-performance ML inference. Microsoft's BrainWave [16] uses FPGA infrastructure to accelerate ML workloads at scale. Even an electric car maker Tesla developed the full self-driving (FSD) chip for autonomous vehicles. There are many start-up companies in this domain. Habana Labs, acquired by Intel in late 2019, developed Gaudi processor for AI training. Graphcore has developed intelligent processing unit (IPU) [17] and deployed in datacenters. Groq's tensor streaming processor [18] optimizes data streaming and computations with fixed task scheduling. Cerabras's wafer-scale engine [19] tries to use a whole wafer as a ML processor to keep a large model without external memories.

1.2   Machine Learning Computations

In this section, we introduce the basic models of deep neural networks (DNNs) and their computations. A DNN model is composed of multiple layers of artificial neurons, where neurons of each layer are inter-connected with the neurons in the neighbor layers. The mathematical model of the neuron comes from Frank Rosenblatt's Perceptron [20] model, as shown in Fig. 1.2. Inspired by the human neuron model, it receives multiple inputs among many input neurons and accumulates their weighted sums with a bias. Then it decides the output through an activation function, where the activation function is non-linear and differentiable, having a step-like characteristic shape. As a result, the output of a neuron is expressed with the following equation:

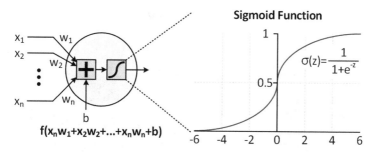

Fig. 1.2 Artificial neuron: perceptron model

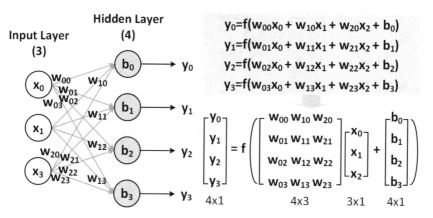

Fig. 1.3 Fully connected layer

$$y = f(w_1 x_1 + w_2 x_2 + \ldots + w_n x_n + b) \qquad (1.1)$$

Based on the network connection, there are three major layers in the DNN models, which determines the actual computations: fully connected, convolutional, and recurrent layer.

1.2.1 Fully Connected Layer

Figure 1.3 shows the fully connected layer that interconnects the neurons in the input layer to the neurons in the next layer. The input vector is the values of input neurons, a 3×1 vector in this case, and the output vector is the values of output neurons, a 4×1 vector. Each connection of the fully connected network between the two layers represents a weight parameter in the model. For example, W_{01} represents a weight parameter of the connection between the input neuron 0 and the output neuron 1. Collectively, the network becomes a 4×3 weight matrix. In addition, each output neuron has a bias, so the layer has 4×1 bias vector. For each output neuron, we can

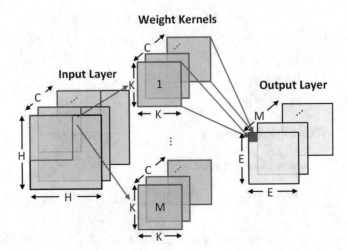

Fig. 1.4 Convolutional layer

write the output value y using Eq. 1.1. As a result, the equations can be formulated into a matrix-vector equation as follows:

$$y = f(Wx + b) \tag{1.2}$$

If the model includes multiple layers, which is the case of deep neural networks, the matrix operations will be cascaded one by one. Traditional multi-layer perceptron (MLP) models as well as the latest transformer models [21] are based on the fully connected layer.

1.2.2 Convolutional Layer

The convolutional layer iteratively performs 3-d convolution operations on the input layer using multiple weight kernels to generate the output layer, as illustrated in Fig. 1.4. The input layer has multiple 2-d input feature maps, sized $H \times H \times C$, and the size of each kernel is $K \times K \times C$. For computation, it performs 3-d convolution operations from top-left to bottom-right for each kernel with a stride of U. A single convolution operation accumulates all the inner products between the input and the kernel. As a result of scanning for a kernel, it gets a single output feature map sized $E \times E$. By repeating this process for all kernels, the convolutional layer produces the final output layer, sized to $E \times E \times M$. The equation for an output point in the convolutional layer is as follows:

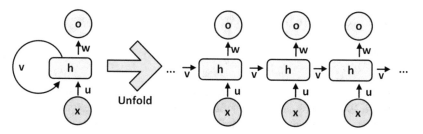

Fig. 1.5 Recurrent layer

$$\mathbf{O}[u][x][y] = \mathbf{B}[u] + \sum_{k=1}^{C-1} \sum_{i=1}^{K-1} \sum_{j=1}^{K-1} \mathbf{I}[k][Ux + i][Uy + j]\mathbf{W}[u][k][i][j],$$

$$(1.3)$$

$$0 \le u < M, 0 \le x, y < E, E = \frac{(H - R + U)}{U}$$

Many convolution neural network (CNN) models use a number of convolutional layers with a few fully connected layers at the end for image classification and recognition task [22].

1.2.3 Recurrent Layer

Figure 1.5 shows the recurrent layer that has a feedback loop from the output to the input layer in the fully connected setting. In this layer, the cell state of the previous timestamp affects the current state. Its computation is also matrix-vector multiplication but involves multiple steps with dependency. Its cell and output value are expressed as follows. The hyperbolic tangent is usually used for activation function in the recurrent layer.

$$h_t = f(U_h x_t + V_h h_{t-1} + b_h), \ O_t = f(W_h h_t + b_o) \qquad (1.4)$$

Recurrent neural networks (RNNs) such as GRUs [23] and LSTMs [24] are based on this type of layer and popularly used for speech recognition.

1.3 von Neumann Bottleneck

1.3.1 Memory Wall Problem

Von Neumann architecture [25] is a computer architecture proposed by John von Neumann in 1945, which broadly consists of a compute unit that executes a program written by a user and a memory unit that stores both the user's program and data required to run the program. Most modern computer systems, including CPU and GPU, fall into this architecture. With the Moore's law that states the number of transistors on a computer chip doubles every 18 months [26] and the process technology scaling, its compute performance has been rapidly improved, as shown in Fig. 1.6. On the other hand, the memory device has been developed to increase its capacity, not the performance. Therefore, the performance gap between the two separated devices gets wider and wider, and it becomes a major performance issue in the system. This memory wall problem causes the data movement issue or limits the memory bandwidth between the compute and memory device. It is often called von Neumann bottleneck because all the computers with von Neumann architecture inevitably have this bottleneck simply because they have separated compute and memory devices.

1.3.2 Latest AI Accelerators with High-Bandwidth Memories

The von Neumann bottleneck has been mitigated with a hierarchical memory structure. Processors include the fastest but smallest SRAM-based cache on-chip to leverage the temporal and spatial locality. Outside of the processor chip, there exists the main memory of the system based on DRAM. DRAM is fast and has a larger capacity than SRAM. After that, the system has solid-state drives (SSD) for high storage capacity. However, as the DNN models get deeper and bigger to the tera-bytes level, the ML workloads require even higher bandwidth between the processor chip and the main memory. Even worse, the process technology scaling

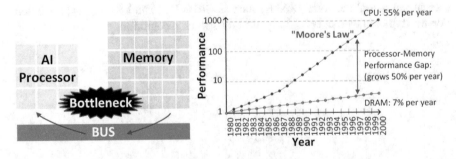

Fig. 1.6 von Neumann Bottleneck and memory wall problem

faces strong challenges with the end of Moore's law below 10 nm technology node [27].

To overcome the von Neumann bottleneck and the slow-down of process scaling, many companies propose array-type architectures to accelerate data-intensive ML processing along with 3-d stacking DRAM technology called high-bandwidth memory (HBM) [28] to provide higher bandwidth between the compute and memory devices. Google developed their own AI chip named TPU to serve the inference and training workloads in datacenters with a better cost and energy efficiency [12]. Intel recently released the NNP-T processor [29] and Habana Labs Gaudi processor [30] for training workloads. Start-up companies such as Graphcore [31] and Groq [18] are also based on this architecture. However, although these AI accelerators with HBM technology can mitigate the bandwidth bottleneck up to a couple TB/s level, they cannot address it eventually as they still fall into von Neumann architecture. In addition, the HBM suffers from high-power dissipation and low capacity [32]. Table 1.1 shows the summary of the hardware specifications of the latest AI accelerators [33].

1.4 Processing-in-Memory Architecture

1.4.1 Paradigm Shift from Compute to Memory

An architectural paradigm called processing-in-memory (PIM) takes an alternative approach to the conventional von Neumann architecture to solve the memory bandwidth problem. It is not a new concept, as it is first introduced in 1970s [34] and has many subsequent works [35, 36]. Recently, PIM has gotten increasing attention amid the memory wall crisis caused by modern ML applications requiring high bandwidths.

In PIM architecture, instead of fetching data from the memory unit to compute unit, data stays in the memory, while the merged logic performs computations in place without moving data outside. As Fig. 1.7 illustrates, this approach is a radical change in the computer architecture; the traditional and near-memory architectures basically have the same memory hierarchy to utilize the external memory bandwidth efficiently, but the PIM merges the compute and memory devices so that it does not have any problems in external data movement. This fundamental shift from compute-centric architecture to memory-centric or combined architecture gets much attention to solve the von Neumann bottleneck, especially for data-intensive applications such as AI and ML. On top of performance improvement, it saves significant energy by replacing expensive external data transfers with on-chip data movements.

Table 1.1 Latest AI accelerators

Metric	Google TPU v3	Nvidia V100	Nvidia A100	Cerebras WSE	GraphCore IPU1	GraphCore IPU2
Technology node	12 nm (16 nm est.)	TSMC 12 nm	TSMC 7 nm	TSMC 16 nm	TSMC 16 nm	TSMC 7 nm
Die area (mm^2)	648 (600 est.)	815	826	46225	900 (est.)	823
Transistor count (B)	11 (est.)	21	54.2	1200	23.6	59.4
Architecture	Systolic array	SIMD + TC	SIMD + TC	MIMD	MIMD	MIMD
Theoretical TFLOPS (16-bit mixed precision)	123	125	312	2500	125	250
Freq (GHz)	0.92	1.5	1.4	Unknown	1.6	Unknown
DRAM capacity (GB)	32	32	80	N/A	N/A	112
DRAM BW (GB/s)	900	900	2039	N/A	N/A	64 (est.)
Total SRAM capacity	32 MB	36 MB (RF+L1+L2)	87 MB (RF+L1+L2)	18 GB	300 MB	900 MB
SRAM BW (TB/s)	Unknown	224 @RF +14 @L1 +3 @L2	608 @RF +19 @L1 +7 @L2	9000	45	47.5
Max TDP (W)	450	450	400	20K	150	150 (est.)

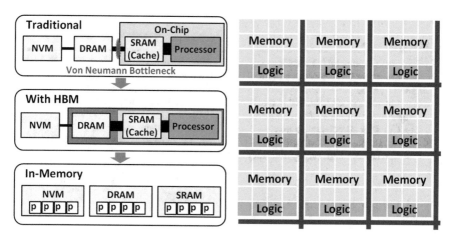

Fig. 1.7 Processing-in-memory architecture

1.4.2 Challenges

Although it looks promising, PIM has many challenges as it needs to integrate logic units into the memory module. The three notable challenges in PIM design are process accessibility, architecting, and designing considering physical constraints, and software stack and usability.

Among many memory technologies, SRAM is the only memory type that we can build using a commercially available logic process. This is why many PIM prototypes are based on SRAM [37–40]. It is possible to fabricate with a logic process and easy to customize both memory cell and peripheral circuits. As the cell size is the biggest among others, SRAM-based PIM has the least area restriction on the logic integration. Except SRAM, DRAM and non-volatile memory (NVM) processes are difficult to access. Memory vendors such as Samsung, SK Hynix, and Micron have their own memory processes, but they are not open to outside. Since the process design kit (PDK) is not accessible, most researchers cannot even simulate the basic circuits. There have been many PIM architecture proposals for DRAM [41, 42]; however, they only evaluate the architectures at a performance simulator level without much physical design. Since the DRAM process is vastly different from the logic process, focusing on increasing cell capacity and cell density, it is hard to convince that the proposed PIM architectures are feasible to be fabricated with only simulations.

It is imperative for chip designers to choose what function they should put into the memory in the PIM design. They cannot implement various functions or too generic logic as the silicon area is limited. In addition, the chip will lose the memory capacity for the area of the logic merged. Another challenge is that the logic design should be physically aligned with the memory cell design to maximize the internal bandwidth.

Table 1.2 PIM opportunities and challenges

PIM opportunities	PIM challenges

1. Non-von Neuman Architecture
  → Can solve von Neumann bottleneck.
2. Converged Logic + Memory
  → Can achieve high internal bandwidth.
3. Suitable for data-intensive workloads
  → Good for AI/ML applications
4. Little external data movement
  → Can achieve high energy efficiency
	1. Process accessibility
	  : Memory process is difficult to use
	2. Limited area resource
	  : What function logic should the designer add?
	3. Physical layout constraint
	  : To maximize the internal compute bandwidth
	4. SW stack for PIM deployment
	  : Revisit a whole SW stack for wide adoption

The software stack is the last hurdle in the PIM design. It is essential for the wide-spreading adoption of PIM as a new device. Unlike traditional memory devices, PIM is not a passive device anymore as it can perform logic operations at the same time. What this means is that we need a fundamental change in the software side either. For real PIM system optimizations, we need to revisit a whole software stack, including programming language, compiler, driver, and run-time. Otherwise, it will not be able to outperform the existing von Neumann computer's performance and usability. Table 1.2 summarizes the opportunities and challenges of PIM technology.

1.5   Book Organization

This book organizes as follows. In Chap. 2, we study the backgrounds of the PIM technology, including basic memory operations of various memories such as SRAM, DRAM, and Resistive RAM (ReRAM). We also discuss the PIM's design constructions and approaches in this chapter. From Chaps. 3–5, we will investigate significant PIM designs in the major memory technologies: SRAM, DRAM, and ReRAM. Each chapter will cover comprehensive design technologies required for PIM, including in-memory circuit processing, memory macro design, data mapping strategy, and architecture. In Chap. 6, we will focus on the PIMs designed for ML training. We will discuss the systems side of PIM, including software and programming interface, in Chap. 7, for the wide adoption of the technology. Finally, we will conclude our book with future remarks in Chap. 8.

References

1. Y. LeCun, Y. Bengio, G. Hinton, Deep learning. Nature **521**(7553), 436–444 (2015)
2. A. Krizhevsky, I. Sutskever, G.E. Hinton, ImageNet classification with deep convolutional neural networks. Adv. Neural Inform. Process. Syst. **25**, 1097–1105 (2012)

3. G. Hinton, L. Deng, D. Yu, G.E. Dahl, A.-r. Mohamed, N. Jaitly, A. Senior, V. Vanhoucke, P. Nguyen, T.N. Sainath, B. Kingsbury, Deep neural networks for acoustic modeling in speech recognition: the shared views of four research groups. IEEE Signal Process. Mag. **29**(6), 82–97 (2012)
4. Y. Goldberg, Neural network methods for natural language processing. Synth. Lect. Hum. Lang. Technol. **10**(1), 1–309 (2017)
5. V. Kepuska, G. Bohouta, Next-generation of virtual personal assistants (Microsoft Cortana, Apple Siri, Amazon Alexa and Google Home), In *2018 IEEE 8th Annual Computing and Communication Workshop and Conference (CCWC)*. IEEE, Piscataway (2018), pp. 99–103
6. M. Teichmann, M. Weber, M. Zoellner, R. Cipolla, R. Urtasun, MultiNet: real-time joint semantic reasoning for autonomous driving, in *2018 IEEE Intelligent Vehicles Symposium (IV)*. IEEE, Piscataway (2018), pp. 1013–1020
7. H.A. Pierson, M.S. Gashler, Deep learning in robotics: a review of recent research. Adv. Robot. **31**(16), 821–835 (2017)
8. M.I. Jordan, T.M. Mitchell, Machine learning: trends, perspectives, and prospects. Science **349**(6245), 255–260 (2015)
9. Y.H. Chen, T. Krishna, J.S. Emer, V. Sze, Eyeriss: an energy-efficient reconfigurable accelerator for deep convolutional neural networks. IEEE J. Solid-State Circuits **52**(1), 127–138 (2016)
10. Z. Yuan, Y. Liu, J. Yue, Y. Yang, J. Wang, X. Feng, J. Zhao, X. Li, H. Yang, STICKER: an energy-efficient multi-sparsity compatible accelerator for convolutional neural networks in 65-nm CMOS. IEEE J. Solid-State Circuits **55**(2), 465–477 (2019)
11. J. Lee, C. Kim, S. Kang, D. Shin, S. Kim, H.J. Yoo, UNPU: A 50.6 TOPS/W unified deep neural network accelerator with 1b-to-16b fully-variable weight bit-precision, in *2018 IEEE International Solid-State Circuits Conference-(ISSCC)*. IEEE, Piscataway (2018), pp. 218–220
12. N.P. Jouppi, C. Young, N. Patil, D. Patterson, G. Agrawal, R. Bajwa, S. Bates, S. Bhatia, N. Boden, A. Borchers, R. Boyle, P.-l. Cantin, C. Chao, C. Clark, J. Coriell, M. Daley, M. Dau, J. Dean, B. Gelb, T.V. Ghaemmaghami, R. Gottipati, W. Gulland, R. Hagmann, C.R. Ho, D. Hogberg, J. Hu, R. Hundt, D. Hurt, J. Ibarz, A. Jaffey, A. Jaworski, A. Kaplan, H. Khaitan, D. Killebrew, A. Koch, N. Kumar, S. Lacy, J. Laudon, J. Law, D. Le, C. Leary, Z. Liu, K. Lucke, A. Lundin, G. MacKean, A. Maggiore, M. Mahony, K. Miller, R. Nagarajan, R. Narayanaswami, R. Ni, K. Nix, T. Norrie, M. Omernick, N. Penukonda, A. Phelps, J. Ross, M. Ross, A. Salek, E. Samadiani, C. Severn, G. Sizikov, M. Snelham, J. Souter, D. Steinberg, A. Swing, M. Tan, G. Thorson, B. Tian, H. Toma, E. Tuttle, V. Vasudevan, R. Walter, W. Wang, E. Wilcox, D.H. Yoon, In-datacenter performance analysis of a tensor processing unit, in *Proceedings of the 44th Annual International Symposium on Computer Architecture* (2017), pp. 1–12
13. G. Batra, Z. Jacobson, S. Madhav, A. Queirolo, N. Santhanam, *Artificial-Intelligence Hardware: New Opportunities for Semiconductor Companies*. McKinsey and Company (2019)
14. M. Abadi, P. Barham, J. Chen, Z. Chen, A. Davis, J. Dean, M. Devin, S. Ghemawat, G. Irving, M. Isard, M. Kudlur, J. Levenberg, R. Monga, S. Moore, D.G. Murray, B. Steiner, P. Tucker, V. Vasudevan, P. Warden, M. Wicke, Y. Yu, X. Zheng, TensorFlow: a system for large-scale machine learning, in *12th USENIX Symposium on Operating Systems Design and Implementation (OSDI 16)* (2016), pp. 265–283
15. Mitchell TM, Machine learning, in Amazon (2017). https://aws.amazon.com/machine-learning/inferentia/. Accessed 21 Oct 2021
16. E. Chung, J. Fowers, K. Ovtcharov, M. Papamichael, A. Caulfield, T. Massengill, M. Liu, D. Lo, S. Alkalay, M. Haselman, M. Abeydeera, L. Adams, H. Angepat, C. Boehn, D. Chiou, O. Firestein, A. Forin, K.S. Gatlin, M. Ghandi, S. Heil, K. Holohan, A. El Husseini, T. Juhasz, K. Kagi, R.K. Kovvuri, S. Lanka, F. van Megen, D. Mukhortov, P. Patel, B. Perez, A.G. Rapsang, S.K. Reinhardt, B.D. Rouhani, A. Sapek, R. Seera, S. Shekar, B. Sridharan, G. Weisz, L. Woods, P.Y. Xiao, D. Zhang, R. Zhao, D. Burger, Serving DNNs in real time at datacenter scale with project brainwave. iEEE Micro **38**(2), 8–20 (2018)
17. Ltd G IPU processors, in: *IPU Processors*. https://www.graphcore.ai/products/ipu. Accessed 21 Oct 2021

18. D. Abts, J. Ross, J. Sparling, M. Wong-VanHaren, M. Baker, T. Hawkins, B. Kurtz, Think fast: a tensor streaming processor (TSP) for accelerating deep learning workloads, in *2020 ACM/IEEE 47th Annual International Symposium on Computer Architecture (ISCA)*. IEEE, Piscataway (2020), pp. 145–158
19. Product—chip. in *Cerebras* (2021). https://cerebras.net/chip/. Accessed 21 Oct 2021
20. F. Rosenblatt, The perceptron: a probabilistic model for information storage and organization in the brain. Psychol. Rev. **65**(6), 386 (1958)
21. A. Vaswani, N. Shazeer, N. Parmar, J. Uszkoreit, L. Jones, A.N. Gomez, Ł. Kaiser, I. Polosukhin, Attention is all you need, in *Advances in Neural Information Processing Systems* (2017), pp. 5998–6008
22. S. Lawrence, C.L. Giles, A.C. Tsoi, A.D. Back, Face recognition: a convolutional neural-network approach. IEEE Trans. Neural Netw. **8**(1), 98–113 (1997)
23. K. Cho, B. Van Merriënboer, C. Gulcehre, D. Bahdanau, F. Bougares, H. Schwenk, Y. Bengio, Learning phrase representations using RNN encoder-decoder for statistical machine translation (2014). arXiv preprint arXiv:1406.1078
24. F.A. Gers, J. Schmidhuber, Recurrent nets that time and count, in *Proceedings of the IEEE-INNS-ENNS International Joint Conference on Neural Networks. IJCNN 2000. Neural Computing: New Challenges and Perspectives for the New Millennium*, vol. 3. IEEE, Piscataway (2000), pp. 189–194
25. J. Von Neumann, First draft of a report on the EDVAC. IEEE Ann. Hist. Comput. **15**(4), 27–75 (1993)
26. G.E. Moore, Cramming more components onto integrated circuits (1965)
27. M.M. Waldrop, The chips are down for Moore's law. Nat. News **530**(7589), 144 (2016)
28. D.U. Lee, K.W. Kim, K.W. Kim, K.S. Lee, S.J. Byeon, J.H. Kim, J.H. Cho, J. Lee, J.H. Chun, 25.2 A 1.2 V 8 Gb 8-channel 128 GB/s high-bandwidth memory (HBM) stacked DRAM with effective microbump I/O test methods using 29 nm process and TSV, in *2014 IEEE International Solid-State Circuits Conference Digest of Technical Papers (ISSCC)*. IEEE, Piscataway (2014), pp. 432–433
29. W. Yang, X. Zhang, Y. Tian, W. Wang, J.H. Xue, Q. Liao, Deep learning for single image super-resolution: a brief review. IEEE Trans. Multimedia **21**(12), 3106–3121 (2019)
30. B. Hickmann, J. Chen, M. Rotzin, A. Yang, M. Urbanski, S. Avancha, Intel Nervana neural network processor-T (NNP-T) fused floating point many-term dot product, in *2020 IEEE 27th Symposium on Computer Arithmetic (ARITH)*. IEEE, Piscataway (2020), pp. 133–136
31. Z. Jia, B. Tillman, M. Maggioni, D.P. Scarpazza, Dissecting the graphcore IPU architecture via microbenchmarking (2019). arXiv preprint arXiv:1912.03413
32. N. Chatterjee, M. O'Connor, D. Lee, D.R. Johnson, S.W. Keckler, M. Rhu, W.J. Dally, Architecting an energy-efficient DRAM system for GPUs, in *2017 IEEE International Symposium on High Performance Computer Architecture (HPCA)*. IEEE, Piscataway (2017), pp. 73–84
33. M. Khairy, TPU vs GPU vs Cerebras vs Graphcore: a fair comparison between ML hardware, in *Medium* (2021). https://khairy2011.medium.com/tpu-vs-gpu-vs-cerebras-vs-graphcore-a-fair-comparison-between-ml-hardware-3f5a19d89e38. Accessed 21 Oct 2021
34. H.S. Stone, A logic-in-memory computer. IEEE Trans. Comput. **100**(1), 73–78 (1970)
35. D.G. Elliott, W.M. Snelgrove, M. Stumm, Computational RAM: a memory-SIMD hybrid and its application to DSP, in *Custom Integrated Circuits Conference*, vol. 30. (1992), pp. 1–30
36. D. Patterson, T. Anderson, N. Cardwell, R. Fromm, K. Keeton, C. Kozyrakis, R. Thomas, K. Yelick, Intelligent RAM (IRAM): chips that remember and compute, in *1997 IEEE International Solids-State Circuits Conference. Digest of Technical Papers*. IEEE, Piscataway (1997), pp. 224–225
37. C. Eckert, X. Wang, J. Wang, A. Subramaniyan, R. Iyer, D. Sylvester, D. Blaauw, R. Das, Neural cache: bit-serial in-cache acceleration of deep neural networks, in *2018 ACM/IEEE 45th Annual International Symposium on Computer Architecture (ISCA)*. IEEE, Piscataway (2018), pp. 383–396
38. D. Fujiki, S. Mahlke, R. Das, Duality cache for data parallel acceleration, in *Proceedings of the 46th International Symposium on Computer Architecture* (2019), pp. 397–410

39. J. Wang, X. Wang, C. Eckert, A. Subramaniyan, R. Das, D. Blaauw, D. Sylvester, 14.2 a compute SRAM with bit-serial integer/floating-point operations for programmable in-memory vector acceleration, in *2019 IEEE International Solid-State Circuits Conference-(ISSCC)*. IEEE, Piscataway (2019), pp. 224–226
40. J.H. Kim, J. Lee, J. Lee, J. Heo, J.Y. Kim, Z-PIM: a sparsity-aware processing-in-memory architecture with fully variable weight bit-precision for energy-efficient deep neural networks. IEEE J. Solid-State Circuits **56**(4), 1093–1104 (2021)
41. S. Li, D. Niu, K.T. Malladi, H. Zheng, B. Brennan, Y. Xie, DRISA: a DRAM-based reconfigurable in-situ accelerator, in *2017 50th Annual IEEE/ACM International Symposium on Microarchitecture (MICRO)*. IEEE, Piscataway (2017), pp. 288–301
42. P. Gu, X. Xie, Y. Ding, G. Chen, W. Zhang, D. Niu, Y. Xie, iPIM: Programmable in-memory image processing accelerator using near-bank architecture, in *2020 ACM/IEEE 47th Annual International Symposium on Computer Architecture (ISCA)*. IEEE (2020), pp. 804–817

Chapter 2
Backgrounds

Chengshuo Yu, Hyunjoon Kim, Bongjin Kim, and Tony Tae-Hyoung Kim

2.1 Basic Memory Operations

Dynamic random-access memory (DRAM) and static random-access memory (SRAM) have played a critical role in modern VLSI systems. Semiconductor technology scaling has increased fabricated memory density and provided higher computing power, which is the main driving force of advancements in electronic systems. However, as the semiconductor technology continues to be scaled, DRAM and SRAM undergo various design challenges such as increased leakage current and smaller sensing margin. Numerous research and development works have been conducted to tackle these issues and provide market-required high performance and low power memory solutions.

Mobile computing devices demand nonvolatile memory solutions, so that key data can be stored even without power supply. FLASH has been developed dramatically because of the explosive growth in the mobile electronics. However, FLASH has been mainly used as storage devices, not as computing devices. In general, it is well-known that flash memory has rewrite endurance of 10^6 times, which is much lower than SRAM and DRAM. FLASH is also inferior to DRAM and SRAM in the write speed and write power. Even though various technologies have been developed to improve the endurance of FLASH and lower power write power consumption, no breakthrough technologies are available that can make flash memory comparable to DRAM and SRAM.

C. Yu · H. Kim · T. T.-H. Kim
Nanyang Technological University, Singapore, Singapore
e-mail: e190026@e.ntu.edu.sg; KIMH0003@e.ntu.edu.sg; thkim@ntu.edu.sg

B. Kim (✉)
University of California Santa Barbara (UCSB), Santa Barbara, CA, USA
e-mail: bongjin@ucsb.edu

© The Author(s), under exclusive license to Springer Nature Switzerland AG 2023
J.-Y. Kim et al. (eds.), *Processing-in-Memory for AI*,
https://doi.org/10.1007/978-3-030-98781-7_2

Table 2.1 Device characteristics of mainstream and emerging memory technologies [1]

	Mainstream memories			Emerging memories			
	SRAM	DRAM	NOR	NAND	MRAM	PCRAM	RRAM
Cell area	$>100F^2$	$6F^2$	$10F^2$	$<4F^2$ (3D)	$6–50F^2$	$4–30F^2$	$4–12F^2$
Multi-bit	1	1	2	3	1	2	2
Voltage	<1 V	<1 V	>10 V	>10 V	<1.5 V	3 V	3 V
Read time	~1 ns	~10 ns	~50 ns	~10 μs	<10 ns	<10 ns	<10 ns
Write time	~1 ns	~10 ns	10 μs to 1 ms	0.1–1 ms	<10 ns	~50 ns	<10 ns
Retention	NA	~64 ms	>10 year	>10 year	>10 year	>10 year	>10 year
Endurance	>1E16	>1E16	>1E5	>1E4	>1E15	>1E9	>1E6–1E12
Write energy (/bit)	~fJ	~100 fJ	~100 pJ	~10 fJ	~0.1 pJ	~10 pJ	~0.1 pJ

F feature size of the lithography

Recently, various resistive nonvolatile memory devices such as magnetic RAM (MRAM), ferroelectric RAM (FeRAM), phase change RAM (PCRAM), and resistive RAM (ReRAM) have been introduced (Table 2.1). Even though they are based on different physical mechanisms, they all realize memory operation by using two different resistance values. Among them, ReRAM has gained high interest because of its simple structure and compatibility to CMOS technology. Besides, ReRAM is more reliable, faster, and consumes lower power than FLASH. Even though ReRAM endurance is still lower than that of DRAM and SRAM, it can be applicable to mobile applications, requiring non-volatility and moderate computing power.

This section introduces the basic operation of DRAM, SRAM, and ReRAM. The basic memory operation will be also applied to processing-in-memory with additional design considerations, which will be discussed in detail in other chapters.

2.1.1 SRAM Basics

Static random-access memory (SRAM) has been used as an embedded memory solution in computing systems because of high performance, robustness, and low fabrication cost [2]. SRAM is faster than DRAM because the cross-coupled inverters in the SRAM cell can generate quicker and larger voltage swings at bitlines. SRAM also receives row address and column address simultaneously while DRAM receives row address and column address separately through the same address pins. Therefore, SRAM shows smaller latency than DRAM. The cross-coupled inverters in SRAM cells also maintain the stored data automatically when wordlines are

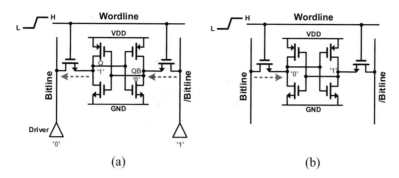

Fig. 2.1 SRAM cell operation: (**a**) write and (**b**) read

turned off. Therefore, SRAM does not require refresh operation and write-back operation. Another key advantage of SRAM is that SRAM is fully compatible with CMOS process technology, which allows SRAM to be easily embedded with computing blocks. However, as CMOS technology scaling continues, SRAM also faces various challenges such as insufficient stability margin, increased leakage current, and difficult supply voltage scaling [2]. Various design techniques have been reported to address these issues [3–11].

Figure 2.1 shows the schematic of the conventional 6T SRAM cell and its write and read operations. The typical SRAM cell consists of six transistors, forming two cross-coupled inverters, and two access transistors. Write operation starts by loading data a bitline pair, followed by turning on a wordline. Then, the data in the bitline pair go to the SRAM cell nodes through the access transistors. For example, as shown in Fig. 2.1a, if the data in Bitline is "0" and the data in /Bitline is "1," Q will be lowered through the access transistor and QB will be raised by /Bitline. Then, the SRAM will store Q = "0." SRAM write operation is mainly limited by the path of writing "0" because the NMOS access transistors can pass low voltage better than high voltage. Therefore, the access transistors need to be stronger than the PMOS transistors to lower Q below the trip point of the inverters in the SRAM cell. SRAM read operation starts by turning on a wordline after precharging bitline pairs. One of the differential bitlines decreases depending on the data stored in the SRAM cell. For example, in Fig. 2.1b, Bitline decreases and /Bitline remains at VDD since Q is "0." A sense amplifier amplifies the differential bitline voltage and generates an output signal.

Figures 2.2 and 2.3 depict a sample SRAM architecture. In general, SRAM consists of an array, row decoding, column multiplexing, sense amplifiers, write drivers, and a controller. During read operation, an accessed cell generates differential voltage at a bitline pair. The differential bitline voltage is connected to a sense amplifier through a column multiplexer. Unlike DRAM, SRAM has sense amplifiers that are shared by multiple columns. Therefore, only one column is connected to a sense amplifier for amplification. No write-back operation is necessary in the unselected columns because SRAM cells can regenerate the stored data through the

Fig. 2.2 SRAM data path

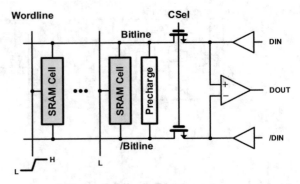

Fig. 2.3 SRAM architecture

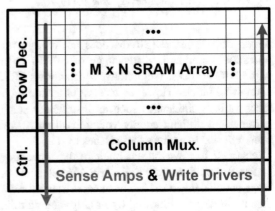

cross-coupled inverters. During write operation, write drivers send the write data to the selected bitlines through the column multiplexer. However, the access transistors in the unselected columns will be on, which can cause unwanted write operation. To mitigate this, the bitlines of unselected columns are precharged to VDD, so that the SRAM cells in the selected row and the unselected columns undergo read operation.

2.1.2 DRAM Basics

Dynamic random-access memory (DRAM) uses a capacitor to store charge for memory operation [12, 13]. DRAM has been popular as a main memory solution because of its compact cell structure and high performance. Figure 2.4 illustrates the unit cell structure of a DRAM cell and its write and read operation principles. A DRAM cell consists of one capacitor for data storage and one transistor for access control. DRAM write operation (Fig. 2.4a) starts by loading data ("1" or "0") into the selected bitline, followed by turning on the selected wordline. Then, the data loaded at the bitline is written into the selected capacitor through charging or discharging. Once the write operation is complete, the selected wordline is disabled,

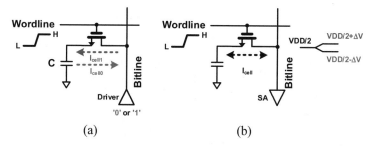

Fig. 2.4 DRAM cell operation: (**a**) write and (**b**) read

and the written data is stored in the capacitor. However, the stored data at the capacitor changes over time because of the leakage current flowing through the access transistor. Multiple leakage paths are formed in the DRAM cell depending on the stored voltage. The stored data will be lost once the stored voltage deviates significantly from the original values and cannot be read through read operation. To tackle this inevitable issue, refresh operation is involved in DRAM to maintain the stored data. Refresh operation read data from selected DRAM cells and write the read data back to the selected DRAM cells using strong "1" and "0." DRAM technologies have developed various techniques such as stacked capacitors [14, 15] and trench capacitors [16, 17] that can achieve the same or larger capacitance after technology scaling.

In DRAM, read operation starts by precharging the bitline with VDD/2. After precharging, the bitline will be floating at VDD/2. Then, the selected wordline is turned on, which will connect the capacitor node to the bitline through the access transistor. The floating bitline voltage will increment or decrement slightly through charge sharing, depending on the cell data. The final voltage formed by the charge sharing can be calculated by the following expression.

$$\text{Final Voltage} = \frac{C_{\text{BL}} \times \frac{VDD}{2} + C_{\text{CELL}} \times V_{\text{CELL}}}{C_{\text{BL}} + C_{\text{CELL}}} \qquad (2.1)$$

Here, C_{BL}, C_{CELL}, V_{CELL} are the bitline capacitance, the cell capacitance, and the cell voltage. Note that the bitline is assumed to be precharged to VDD/2. In general, C_{BL} is much larger than C_{CELL} because a few hundred cells share a bitline. Therefore, the number of cells per bitline should be determined carefully after considering the minimum voltage swing requirement for reliable sensing.

Figure 2.5 depicts a sample DRAM data path. Before read operation, the equalizer precharges the bitline pair with VDD/2. In read operation, a DRAM cell increments or decrements the bitline voltage, and the bitline voltage is amplified by a sense amplifier in each bitline. The amplified voltage will be transferred to the output (DOUT) through the selection signal (BSel). The read operation is destructive since the cell voltage after charge sharing will become Eq. (2.1).

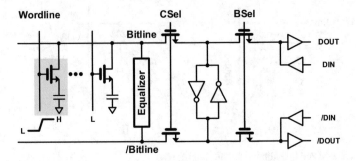

Fig. 2.5 DRAM data path

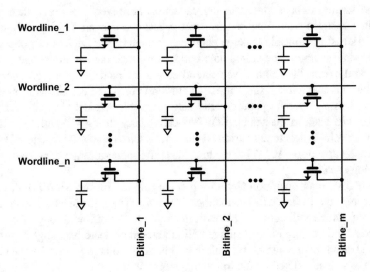

Fig. 2.6 DRAM array architecture

Therefore, the amplified voltage will also be written back to the selected DRAM cell through the bitline to maintain the data.

A sample DRAM architecture is shown in Fig. 2.6. When a row is selected for read operation by turning on a wordline, all the DRAM cells in the selected row will generate voltage increment or decrement in the bitlines. The sense amplifier in each bitline will amplify the small voltage change and generate read data. The read data generated in all the columns cannot be sent to the output ports over one clock cycle because of the limited data width. Therefore, it is necessary to use multiple cycles for reading them out. If not, only a part of the read data will be sent to the outputs. Writing operation also happens in the selected columns. However, the access transistors in the unselected columns will be also on, generating voltage increment or decrement in the bitlines. Therefore, it is necessary to activate the sense

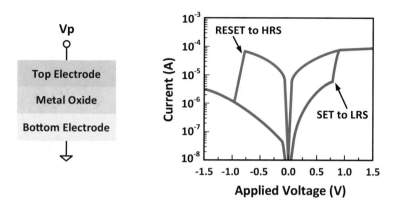

Fig. 2.7 ReRAM device structure and I-V curve [22]

amplifiers and write the strong data after amplification back to the corresponding
DRAM cells.

2.1.3 ReRAM Basics

Resistive memory (ReRAM) is a promising low-power nonvolatile memory solution for various emerging applications such as Internet-of-Things (IoT), wearable
devices, and biomedical devices where batteries or energy harvesting devices supply
power. ReRAM utilizes two distinctive resistance values (HRS: high-resistance state
and LRS: low-resistance state) or more as data storage [18–21]. Figure 2.7 illustrates
a sample metal–insulator–metal ReRAM device structure and its I–V curve. As
shown in Fig. 2.7, the ReRAM device is a two-terminal device including a thick
oxide layer inserted between electrodes. The device switching between HRS and
LRS is controlled by the magnitude and the direction of the applied bias. Typically,
the switching from HRS to LRS is called SET, while the opposite switching is
named RESET. Before setting and resetting ReRAM, one additional operation is
necessary for actual scenarios called FORMING. After fabricating ReRAM devices,
they have very high resistance because of the oxide layer in the ReRAM devices. A
predefined high voltage is applied to the fresh ReRAM devices to form filaments in
the oxide layer. This step is necessary only one time per ReRAM device and called
FORMING. After FORMING, the ReRAM devices are in the HRS state and ready
to be SET. Even if the ReRAM device states are represented by HRS and LRS, the
actual resistance is affected by the applied bias as shown in Fig. 2.7 [22].

   Various ReRAM cells have been reported depending on the required features
[23–30]. One of the most common ReRAM cells is the one-transistor and one-
resistor (1T1R) structure. Figure 2.8 explains the basic 1T1R ReRAM operations
such as SET, RESET, and read operation. SET requires V_{SET} at the bitline (BL),

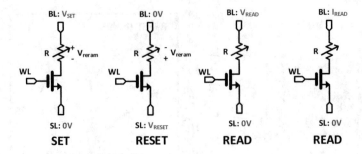

Fig. 2.8 SET, RESET, and read operation of 1T1R ReRAM cell

GND at the source line (SL). Turning on the NMOS access transistor will make current from BL to SL through the ReRAM device. This current will change the ReRAM state from HRS to LRS. One important parameter to consider is the ReRAM set voltage ($V_{R\text{-SET}}$), which is the minimum required voltage across the ReRAM device for SET. When current flows the 1T1R cell, the drain node of the access transistor goes up. Therefore, only a part of V_{SET} is observed across the ReRAM device (V_{reram}). For proper SET, V_{reram} should be larger than $V_{R\text{-SET}}$. RESET occurs when BL is grounded and V_{RESET} is applied to SL. In this case, current flows in the opposite direction of the SET operation, and the ReRAM state switches from LRS to HRS. Like SET, V_{reram} should be larger than the ReRAM reset voltage ($V_{R\text{-RESET}}$). However, the NMOS access transistor has larger voltage drop and V_{reram} of RESET cab be smaller than V_{reram} of SET when the same voltage is used at WL. Therefore, it is necessary to employ a technique that can reduce the voltage drop through the access transistor smaller. Boosted WL voltage can improve the voltage drop at the cost of reliability degradation. ReRAM read operation can be implemented in two different modes (i.e., voltage mode and current mode). In the voltage mode, BL is precharged to read voltage (V_{READ}) and is discharged with different rates based on the ReRAM state. A sense amplifier detects the discharging rates and produces an output signal. V_{READ} needs to be small, so that the ReRAM state is not disturbed. Since the bias condition of the ReRAM read operation is similar to SET, V_{READ} needs to be much smaller than V_{SET} to avoid unwanted SET. In the current mode, predefined read current (I_{READ}) is supplied to the selected ReRAM cell. This current will generate voltages utilizing the ReRAM states, which are sensed by amplifier. The current needs to be small enough to maintain the voltage across the ReRAM device smaller than $V_{R\text{-SET}}$ with enough margins. It is also necessary to regulate the voltage at BL below a certain level to prevent ReRAM state disturbance.

Figure 2.9 describes a sample 1T1R ReRAM array architecture. A row is selected by applying high voltage to the selected wordline. During SET and RESET, the bitlines and the source lines need to be biased properly. The ReRAM architecture in Fig. 2.9 cannot execute SET and RESET at the same time in the selected row. In the SET operation, some ReRAM cells can be set while the others remain unchanged.

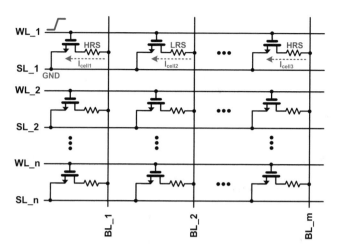

Fig. 2.9 1T1R ReRAM array architecture

Similarly, some ReRAM cells will be reset in the RESET operation while the rest will remain unchanged. This occurs because each source line is shared by a row. SET and RESET can be executed simultaneously when the source lines run vertically like the bitlines. Here, the bitline and the source line of each column can be controlled independently, which facilitates SET and RESET over one cycle. However, this architecture consumes more power than the architecture in Fig. 2.9. Therefore, the ReRAM array architecture needs to be selected carefully based on the system requirement. In the read operation, current will flow from each bitline to the selected source line as shown in Fig. 2.9. The ReRAM states will determine the magnitude of the current in each bitline. For example, I_{cell1} will be smaller than I_{cell2} in Fig. 2.9 since I_{cell1} and I_{cell2} are generated by HRS and LRS, respectively. The current in each bitline will be compared with the average value of I_{cell1} and I_{cell2} by a sense amplifier. The 1T1R ReRAM architecture have faced various challenges in the point of scaling. First, the resistance of the ReRAM devices should be properly defined so that the voltage across the ReRAM devices is large enough to set or reset. If the ReRAM resistance is too low, the voltage drop across the access transistor becomes large. This requires boosted wordline voltage, which will deteriorate the ReRAM device reliability. The lowest ReRAM resistance value can be determined by RESET. Another challenge is ReRAM device parameter scaling. It is well known that ReRAM programming current does not show good scalability. ReRAM set and reset voltages also need to be scaled in a similar rate of CMOS scaling. If ReRAM device parameters are not salable, additional circuit techniques should address the scalability issues. Figure 2.10 summarizes LRS, HRS, ReRAM set voltage, and ReRAM reset voltage in literature. It is obvious that ReRAM set and reset voltages are still too high when considering the supply voltage levels of the mainstream CMOS technologies. They must be scaled below 1 V, so that they can be integrated with advanced CMOS technologies.

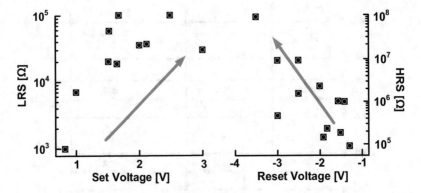

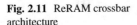

Fig. 2.10 Literature survey of LRS, HRS, ReRAM set voltage, and ReRAM reset voltage [31]

Fig. 2.11 ReRAM crossbar
architecture

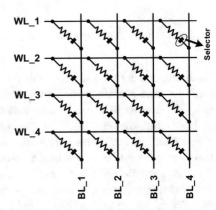

Another popular ReRAM array is the crossbar architecture where ReRAM cells
without access transistors are sandwiched between rows and columns as illustrated
in Fig. 2.11 [32–34]. Since no transistor is used, the crossbar architecture provides
higher area efficiency compared to the 1T1R architecture. The ReRAM cell for
the crossbar architecture includes a selector device to cut the sneak current in the
unselected ReRAM cells. Programming and reading in the crossbar architecture
require more careful design considerations because of the sneak paths formed by
unselected ReRAM cells.

Figure 2.12a, b show two popular ReRAM programming schemes in the crossbar
architecture. The selected row and the selected column are biased with the writing
voltage (VDD in Fig. 2.12) and GND or vice versa relying on the intended programming data. Programming current will flow through the selected cell switching
the resistance state. However, there are additional current paths whose current (i.e.,
sneak current) is not negligible. The additional current paths, named sneak paths,
can vary when employing different write schemes. In the VDD/2 writing scheme
(Fig. 2.12a), the unselected cells in the selected row and the unselected cells in the
selected column generate the sneak current. However, in the VDD/3 write scheme

Fig. 2.12 ReRAM operation
in the crossbar: (**a**) write
using VDD/2, (**b**) write using
VDD/3, and (**c**) read
operation

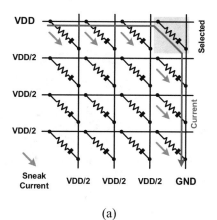

(a)

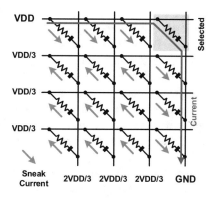

(b)

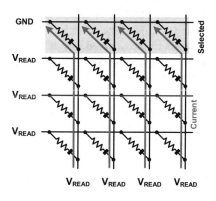

(c)

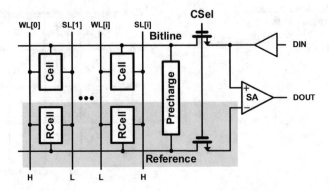

Fig. 2.13 ReRAM data path

(Fig. 2.12b), all the unselected cells will contribute to the sneak current. However, the total current required in each row driver will be less than that of the VDD/2 write scheme. The VDD/3 write scheme has VDD/3 as the potential difference across the unselected cells in the selected row, while the VDD/2 write scheme has VDD/2 for the same cells. The total amount of the sneak current is also data-dependent. The programming current and the sneak current should be provided by the wordline driver in each row. Therefore, the actual array size and the number of programmed cells per cycle needs to be decided carefully after considering the driving capability and the area of the driver. Figure 2.12c explains the read operation in the crossbar architecture. GND is applied to the selected row, and the rest signals are connected to read voltage (V_{READ}). Read current will flow from the bitlines to the grounded selected row through the selected cells. In principle, all the data in the selected row can be read out, which requires a sense amplifier in each column. In actual scenarios, only a part of the row data will be transferred to sense amplifiers. However, all the columns will still consume read current like SRAM.

Figure 2.13 shows a sample data path of ReRAM. When a wordline is selected, the cells in the selected row will generate bitline voltages. Multiple columns share a sense amplifier (SA), so a column multiplexer (CSel in Fig. 2.13) will connect the bitline voltage of the selected column to the sense amplifier for further amplification. Sense amplifiers need a reference voltage for comparison. The reference voltage can be provided through an external signal after testing the fabricated ReRAM. However, this requires comprehensive and time-consuming test sequences, which is not practical. As shown in Fig. 2.13, it is more desirable to implement on-chip reference voltage. One way of generating on-chip reference voltage is using ReRAM replicas. In Fig. 2.13, the Replica ReRAM cell (RCell) is programmed to make the reference voltage higher than the bitline voltage for LRS and lower than the bitline voltage for HRS. The programming of RCell can be designed in various ways. One common way is to use two RCells connected in parallel. One RCell is programmed with HRS, and the other RCell is programmed with LRS. In this case, the equivalent resistance becomes as follows.

$$R_{\text{REFERENCE}} = \frac{R_{\text{HRS}} \times R_{\text{LRS}}}{R_{\text{HRS}} + R_{\text{LRS}}} \qquad (2.2)$$

Another way is to use one RCell and program it with $(R_{\text{HRS}} + R_{\text{LRS}})/2$. However, this requires a complicated control on the programming voltage and the pulse width for accurate programming. Therefore, it is more desirable to use multiple RCells programmed with either HRS or LRS.

2.2 PIM Fundamentals

Processing-in-memory (PIM) has recently attracted significant attention as an alternative computer architecture for the energy-efficient processing of massively parallel arithmetic operations, enabling artificial intelligence and machine learning, especially for battery-operated edge computing devices. Multiply-and-accumulate (MAC) is a critical operation for processing artificial neural networks in edge devices. For instance, a convolutional neural network (CNN) typically requires billions of MAC operations to process a single image classification. Hence, the design of MAC operation units plays a critical role in the overall performance and energy consumption of the hardware accelerator based on PIM architecture.

Emerging memory devices, including a resistive random access memory (ReRAM), a magneto resistive (MRAM), and a phase-change random access memory (PCRAM), are possible candidates for the PIM implementation. They are compact, and hence, they can achieve a high storage capacity and massively parallel MAC operations. While the compact emerging memories are gaining significant attractions for the PIM implementation, they are still not mature yet, and the cost is too high. Hence, a traditional static random-access memory (SRAM) has been most frequently used. Though it has a relatively large bitcell size, it has many other advantages over the emerging candidates, such as scalability, logic compatibility, low cost, and reliability.

A PIM macro can be implemented by reusing a classical two-dimensional array of different cells (SRAM/DRAM/ReRAM), as shown in Fig. 2.14. Here, the MAC operation of standard SRAM is described deeply as an example. A standard six-transistor (6T) SRAM cell is used as a binary PIM unit to perform multiply and accumulate operations. A binary input (i.e., 0 or +1) is applied to each macro row and used as a multiplicand for all the SRAM cells in the same row. A binary weight (i.e., −1 or +1) is stored in an SRAM cell, and then, it is multiplied by the input applied to its wordline (WL). The accumulation is performed column-by-column, and the accumulated result is a voltage difference between a bitline

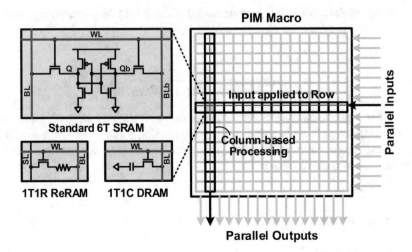

Fig. 2.14 PIM macro using common memory cells (standard 6T SRAM/1T1C DRAM/1T1R ReRAM)

(BL) and a bitline-bar (BLb). A '0' input is represented by a WL high voltage, which creates a discharging path from BL (or BLb) to the ground via an SRAM internal node Q (or Qb). Note that all the bitlines (BLs and BLbs) are precharged to high before the MAC operation is performed. Ideally, all the inputs are applied, and the outputs are generated in parallel; hence, massively parallel binary SRAM PIM operations are performed to maximize throughput and minimize latency. In practice, the essential data conversion for input (digital-to-analog) and output (analog-to-digital) eventually determines the overall performance of the designed PIM macro.

Using the PIM macro in Fig. 2.14, we can map and process the essential arithmetic operations of a fully connected layer of deep neural networks (DNNs). Figure 2.15a shows a pair of binary input and weight mapped to an SRAM cell and an input pair. Binary multiplication is performed in the SRAM cell, and it results in a unit analog accumulation as a voltage difference in the vertical bitlines (BL and BLb). A group of input and weight pairs forms a dot-product, as shown in Fig. 2.15b, left. The dot-product is mapped to a column of the PIM macro, as shown in Fig. 2.15b, right. The unit voltage differences from SRAM cells accumulate in the column, which shares a BL and BLb pair. Finally, a vector-matrix multiplication (i.e., a fully connected layer itself) is mapped to the entire PIM macro, where all the multiplications and accumulates are performed in parallel, enabling massive parallelism, as depicted in Fig. 2.15c.

The PIM macro can assign a convolutional layer by unrolling and mapping high dimensional filter weights and input feature maps into the PIM macro. Figure 2.16a shows the mapping of a pair of input and weight from a convolutional layer configuration into the macro. A two-dimensional (2D) filter weights are unrolled and mapped into a column of four bitcells, as shown in Fig. 2.16b. A three-dimensional

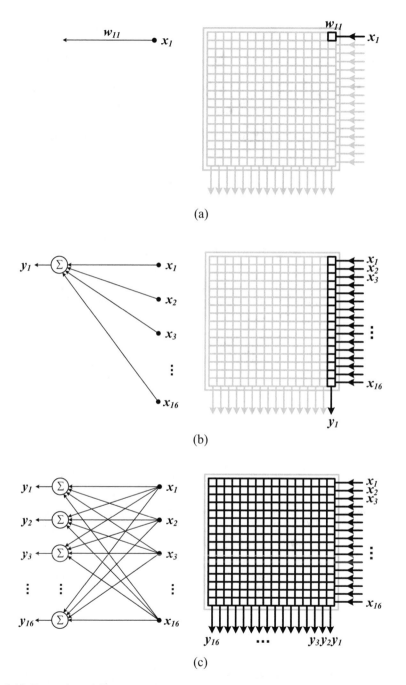

Fig. 2.15 Processing a fully connected layer using an SRAM-based PIM macro: (**a**) a multiplication; (**b**) a dot-product; (**c**) a vector-matrix multiplication

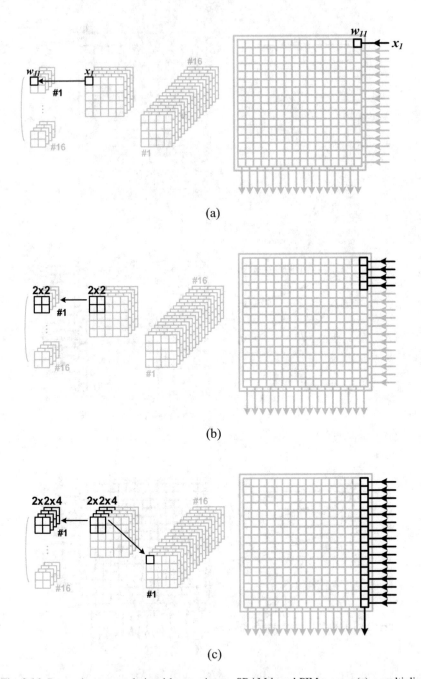

Fig. 2.16 Processing a convolutional layer using an SRAM-based PIM macro: (**a**) a multiplication; (**b**) a dot-product for a 2D filter; (**c**) a dot-product for a 3D filter; (**d**) a vector-matrix for a 4D filter; (**e**) after 16 cycles of vector-matrix operations

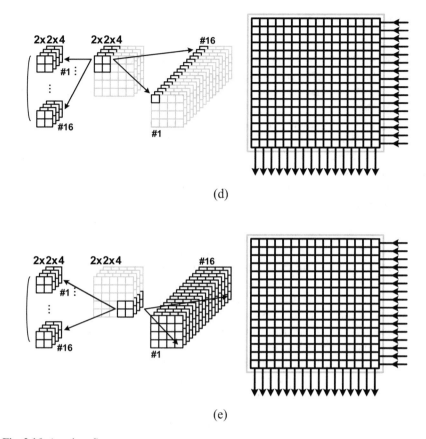

(d)

(e)

Fig. 2.16 (continued)

(3D) filter weights and input feature maps, composed of multiple channels of 2D filters and input feature maps, are mapped to the entire column of the PIM macro, as depicted in Fig. 2.16c. Note that the 3D filter and input feature map can be mapped to multiple macro columns when the number of filter and input feature map element pairs is larger than the number of bitcells in a single macro column. Figure 2.16d expands one more dimension of the convolutional layer processing (i.e., output channels or the channels of 3D filters), which can be processed in parallel using multiple columns in the PIM macro. Each column output corresponds to a pixel of each 2D output feature map. To complete the 3D output feature map generation, we will reuse the same PIM macro and process while sliding the window of the input feature map and complete the 3D output feature map, as illustrated in Fig. 2.16e.

2.3   PIM Output Read-out

Input and output delivery of the PIM macro add extra operation latency and energy consumption. Although I/O and periphery blocks in many state-of-the-art macro designs are not emphasized, it is important to note their key component and their associated challenges.

Among I/O and periphery blocks, a sense amplifier (SA) is one of the most critical components in many SRAM based PIM macro designs that utilize bitline discharge for MAC operation. Figure 2.17a is the standard latch-type SA with the minimum number of transistors. However, the voltage drop across the pass transistor causes limitation in input voltages, ultimately reducing the noise margin and the voltage swing of the bitline. Figure 2.17b is an improved version of the StrongARM Latch [35] to achieve low static power, produce higher output dynamic range, and minimize the offset caused by the input differential pair. However, the circuit operation phases include voltage gain, which introduces an amplified offset issue from the VN and VP transistor mismatch. The separate SR latch added to the right part of Fig. 2.17b is one of the techniques to cancel the offset by establishing different discharge rates from the input pair. Figure 2.17c also provides offset cancellation from programmable capacitors. However, to control the random offset values, the number of capacitor switches increases and ultimately raises power dissipation and lowers operation speed.

In many cases, SA output requires analog-to-digital (A2D) conversion as the function of SA is an analog comparator. Thus, ADC is another critical block in PIM macros that can directly affect the performance of the PIM macro.

Figure 2.18a describes the operating principle of a single-slope ADC operation [37]. For illustration purposes, a column of bitcells is simplified to have 13 PIM bitcells for a nine-input dot-product and a 2-bit single-slope column ADC. Thus, an N-bit single-slope ADC requires 2^N-1 cycles to complete single data conversion.

The operating principle of the binary-searching ADC [38] is shown in Fig. 2.18b. The top 384 bitcells are simplified to a black box with a fixed dot-product result (+27), and the following 96 bitcells are separated into two grounds for representing weight "+1" (white boxes) and weight "−1" (gray boxes) when operating as the binary-searching ADC. Each group has four input signals via RWLs that control 6, 6, 12, and 24 bitcells, respectively. The binary-searching ADC takes five cycles to generate a 5-bit output code, as shown at the bottom of Fig. 2.18b.

The 4-bit flash ADC [36] has advantages in power consumption, performance, and area tradeoff. Moreover, the 15 reference voltage levels of the flash ADC can be easily adjusted by changing the voltage input of the resister diode ladder, so that the read bitline (RBL) dynamic range can also be tunned easily. Note that the Flash ADC comprises clocked comparators that can be readily implemented in the column-pitch of the PIM macro. The connection of 15 comparators to 15 read bitlines (RBLs) to serve as part of the computation caps is described in detail in Fig. 2.18c.

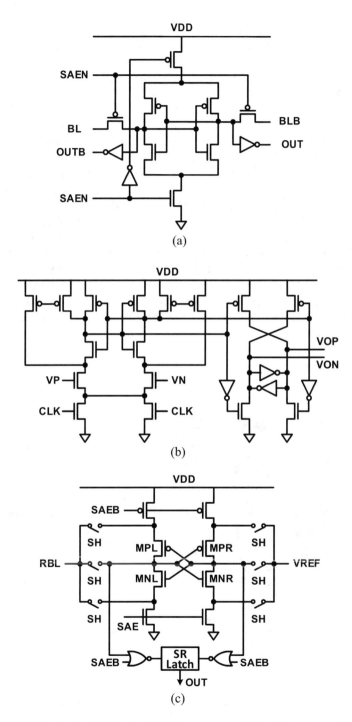

Fig. 2.17 Detailed circuit of (**a**) standard latch-type sense amplifier, (**b**) an improved version of the StrongARM latch [35], and (**c**) an offset cancellation latch [36]

Charge-sharing ADC (CSH_ADC) shown in Fig. 2.18d is a serial integrating ADC that can benefit from the narrow distribution ADC output code. In a PIM macro architecture where energy and area efficiencies are prioritized over other performance metrics, ADC architecture such as area-intensive SAR and power-hungry flash ADCs are avoided. Biswas and Chandrakasan [39] claim that CSH_ADC is suitable for the design, as the macro showed a narrow output distribution in its partial CNN output, which is translated to low operation cycles that do not extend to performance loss even with the serial nature of the ADC. The consisting blocks in CSH_ADC can be broken into three parts: an integrator, an SA, and a logic block. The capacitive charge-sharing aspect of the integrator is implemented by the replica bitlines from the PIM cell, which can also be beneficial to track the variation in bitline capacitance. The SA is a standard StrongARM latch-type architecture with a PMOS input pair due to the expected common mode voltage levels of V_p and V_n being closer to GND rail. The final logic block provides control signals PCH_R, EQ_p, EQ_n, and SA_EN using the global timing signals ϕ_1 and ϕ_2. ADC begins its conversion process by first determining the sign of the output from the two input nodes and followed by the replica bit-line precharge for equalize/integrate operation. Integrate operation repeats until the SA output is flipped to flag the end-of-conversion (EOC), and finally, the counter counts the equalize pulses (EQ_n) to generate the digital value.

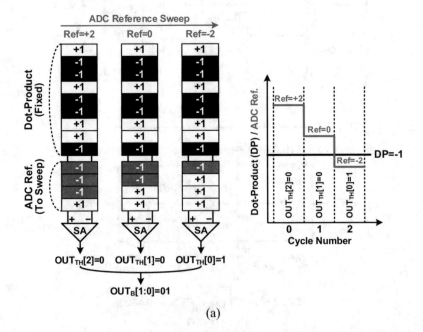

(a)

Fig. 2.18 Detailed circuit of (**a**) single-slope column ADC [37]; (**b**) binary-searching column ADC [38]; (**c**) 4-bit flash ADC [36]; (**d**) charge-sharing ADC [39]

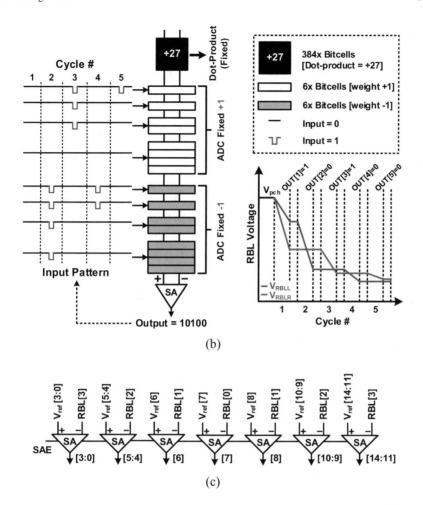

(b)

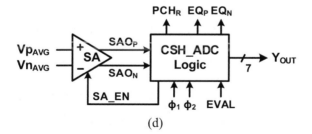

(c)

(d)

Fig. 2.18 (continued)

2.4   PIM Design Challenges

While the analog PIM macros present outstanding efficiency numbers, serious
design challenges exist. Most well-known issues are process, temperature, and
voltage (PVT) variation induced computation nonlinearity and DAC/ADC overhead.

Figure 2.19a depicts input offset error of analog circuits in PIM macro (i.e.,
bitcell, SA, and ADC) caused by process variation. Figure 2.19a, left, illustrates the
error distribution of the output ADC code for identical MAC operations. Although
the memory bitcells in the PIM array have a regular structure, the difference in
MAC results exists due to process variation in the fabrication of memory bitcells.
The variation of one single bitcells and a whole column are shown in Fig. 2.19a
right. Figure 2.19a top right describes the distribution of discharge current when
one bitcell processes multiplicate operation using current discharge, and Fig. 2.19a
bottom right shows the bitline (BL) voltage allocation after completing the dot-
product operation in one column-based neuron. Overall, the process variation
fluctuates the bitline voltage representing the dot-product result and increases the
possibility of producing incorrect output ADC code. In terms of neural networks,
the generated output code becomes the new input activation for the next layer and
used to calculate another dot-product. Thus, the wrong output code of one layer
propagates through several computations and generates classification error at the
end, reducing the application task's accuracy.

The computation nonlinearity happens when more rows are activated in parallel
to improve the computation efficiency, as shown in Fig. 2.19b. The bitline voltage
representing dot-product results decrease when more "1"s in the column are
added and cause a dynamic range limit. The accumulation linearity is significantly
degraded if the bitline voltage drops too low, as shown by the red dot line of Fig.
2.19b right.

The overhead of digital-to-analog and analog-to-digital converter (DAC/ADC)
for data transmission is also a significant concern for PIM macro. As shown in Fig.
2.19c, the DAC/ADC not only spends huge circuit area and energy consumption but
also increases the latency of the neural network accelerator. In addition, the typical
ADC has the fixed bit-precision, resulting in limitation about reconfigurability.

On the other hand, digital PIM macros suffer from different critical issues: low
area efficiency and high-power consumption. Figure 2.20a describes the modern
neural network accelerator containing a complete digital process elements (PEs)
array that processes massive MAC operation synchronously [41]. With the help of
hierarchical memory and data reuse strategy, this work improves computation efficiency while also saving energy since memory access energy exceeds energy from
MAC operations. Figure 2.20b illustrates one PIM column with the parallel adder
tree, which performs massively parallel accumulation operation without additional
registers to store input activations and partial sums [40]. Note that the bit-serial
multiplication also improves energy efficiency in the tradeoff of operation latency.
In addition, the entire digital approach avoids the compute nonlinearity and poor
scaling of analog circuits. However, the full digital PE comprises more arithmetic
circuits, which not only occupies large area but also costs larger static/dynamic
energy compared to the bitcell of analog PIM.

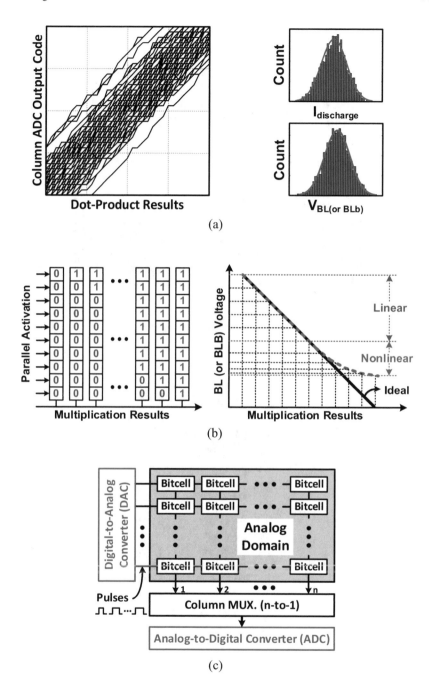

Fig. 2.19 Challenges of analog PIM macro: (**a**) process variation; (**b**) nonlinearity; (**c**) ADC overhead

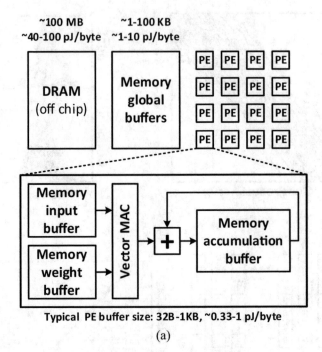

Typical PE buffer size: 32B-1KB, ~0.33-1 pJ/byte

(a)

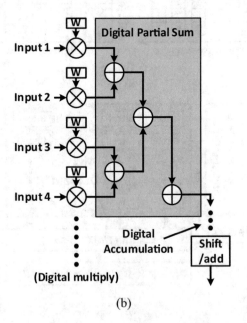

(b)

Fig. 2.20 (a) Simplified block diagram of a typical digital DNN accelerator; (b) A column-based dot-product circuit using digital PIM [40]

References

1. S. Yu, P.-Y. Chen, Emerging memory technologies: Recent trends and prospects. IEEE Solid-State Circuits Magaz. **8**(2), 43–56 (2016)
2. K. Zhang, F1: Embedded memory design for nano-scale VLSI systems, in *2008 IEEE international solid-state circuits conference—Digest of technical papers* (2008), pp. 650–651
3. N. Shibata et al., 1-V 100-MHz embedded SRAM techniques for battery-operated MTC-MOS/SIMOX ASICs. IEEE J. Solid State Circuits **35**(10), 1396–1407 (2000)
4. K. Agawa et al., A bitline leakage compensation scheme for low-voltage SRAMs. IEEE J. Solid State Circuits **36**(5), 726–734 (2001)
5. K. Nii et al., A 90-nm low-power 32-kB embedded SRAM with gate leakage suppression circuit for mobile applications. IEEE J. Solid State Circuits **39**(4), 684–693 (2004)
6. T.-H. Kim et al., A 0.2V, 480kb subthreshold SRAM with 1k cells per bitline for ultra-low voltage computing. IEEE J. Solid State Circuits **43**(2), 518–529 (2008)
7. T.-H. Kim et al., A voltage scalable 0.26V, 64kb 8T SRAM with Vmin lowering techniques and deep sleep mode. IEEE J. Solid State Circuits **44**(6), 1785–1795 (2009)
8. T. Kim et al., Design of a temperature-aware low voltage SRAM with self-adjustable sensing margin enhancement for high temperature applications up to 300°C. IEEE J. Solid State Circuits **49**(11), 2534–2546 (2014)
9. B. Wang et al., Design of an ultra-low voltage 9T SRAM with equalized bitline leakage and CAM-assisted energy efficiency improvement. IEEE Trans. Circuits Syst. TCAS-I Regul. Pap. **62**(2), 441–448 (2015)
10. A. Do et al., 0.2 V 8T SRAM with PVT-aware bit-line sensing and column-based data randomization. IEEE J. Solid State Circuits **51**(6), 1487–1498 (2016)
11. C. Duan et al., Energy-efficient reconfigurable SRAM: Reducing read power through data statistics. IEEE J. Solid State Circuits **52**(10), 2703–2711 (2017)
12. T. Kirihata et al., An 800 MHz embedded DRAM with a concurrent refresh mode, in *IEEE int. solid-state circuits conference (ISSCC)*, (IEEE, Piscataway, 2004), pp. 206–523
13. M. Kumar et al., A simple and high-performance 130 nm SOI EDRAM technology using floating-body pass-gate transistor in trench-capacitor cell for system-on-a-chip (SoC) applications, in *IEEE int. electron devices meeting (IEDM)*, (IEEE, Piscataway, 2003), pp. 17.4.1–17.4.4
14. S. Yamamichi et al., A stacked capacitor technology with ECR plasma MOCVD (Ba,Sr)TiO/sub 3/and RuO/sub 2//Ru/TiN/TiSi/sub x/ storage nodes for Gb-scale DRAMs. IEEE Trans. Electron Devices **44**(7), 1076–1083 (1997)
15. S. Yamamichi et al., An ECR MOCVD (Ba,Sr)TiO/sub 3/based stacked capacitor technology with RuO/sub 2//Ru/TiN/TiSi/sub x/storage nodes for Gbit-scale DRAMs, in *IEEE int. electron devices meeting (IEDM)*, (IEEE, Piscataway, 1995), pp. 119–122
16. G. Aichmayr et al., Carbon/high-k trench capacitor for the 40nm DRAM generation, in *IEEE symp. on VLSI technology*, (IEEE, Piscataway, 2007), pp. 186–187
17. T.S. Boscke et al., Tetragonal phase stabilization by doping as an enabler of thermally stable HfO2 based MIM and MIS capacitors for sub 50nm deep trench DRAM. Int. Electron Devices Meet. **2006**, 1–4 (2006)
18. H.-S.P. Wong et al., Metal-oxide RRAM. Proc. IEEE **100**(6), 1951–1970 (2012)
19. M.-F. Chang et al., Low VDDmin swing-sample-and-couple sense amplifier and energy-efficient self-boost-write-termination scheme for embedded ReRAM macros against resistance and switch-time variations. IEEE J. Solid State Circuits **50**(11), 2786–2795 (2015)
20. S. Zuloaga et al., Scaling 2-layer RRAM cross-point array towards 10 nm node: A device-circuit co-design, in *IEEE int. symp. on circuits and systems (ISCAS)*, (2015), pp. 193–196
21. A. Bricalli et al., SiOx-based resistive switching memory (RRAM) for crossbar storage/select elements with high on/off ratio, in *IEEE int. electron devices meeting (IEDM)*, (2016), pp. 4.3.1–4.3.4

22. Y. Chen et al., Reconfigurable 2T2R ReRAM architecture for versatile data storage and computing in-memory. IEEE Trans. VLSI Syst. **28**(12), 2636–2649 (2020)
23. H.Y. Lee et al., Low power and high speed bipolar switching with a thin reactive Ti buffer layer in robust HfO2 based RRAM, in *IEEE int. electron devices meeting (IEDM)*, (IEEE, Piscataway, 2008), pp. 1–4
24. C. Zambelli et al., Electrical characterization of read window in ReRAM arrays under different SET/RESET cycling conditions, in *IEEE 6th int. memory workshop*, (IEEE, Piscataway, 2014), pp. 1–4
25. E. Vianello et al., Resistive memories for ultra-low-power embedded computing design, in *IEEE int. electron devices meeting (IEDM)*, (IEEE, Piscataway, 2014), pp. 6.3.1–6.3.4
26. A. Fantini et al., Intrinsic program instability in HfO2 RRAM and consequences on program algorithms, in *IEEE int. electron devices meeting (IEDM)*, (IEEE, Piscataway, 2015), pp. 7.5.1–7.5.4
27. Z.-Q. Wang et al., Cycling-induced degradation of metal-oxide resistive switching memory (RRAM), in *IEEE int. electron devices meeting (IEMD)*, (IEEE, Piscataway, 2015), pp. 7.6.1–7.6.4
28. H.B. Lv et al., BEOL based RRAM with one extra-mask for low cost, highly reliable embedded application in 28nm node and beyond, in *IEEE int. electron devices meeting (IEMD)*, (IEEE, Piscataway, 2017), pp. 2.4.1–2.4.4
29. P.-Y. Chen et al., Design tradeoffs of vertical RRAM-based 3-D cross-point array. IEEE Trans. Very Large-Scale Integr. Syst. **24**(12), 3460–3467 (2016)
30. P.-Y. Chen, Compact modeling of RRAM devices and its applications in 1T1R and 1S1R array design. IEEE Trans. Electron Devices **62**(12), 4022–4028 (2015)
31. L. Lu et al., ReRAM device and circuit co-design challenges in nano-scale CMOS technology, in *16th IEEE Asia Pacific conference on circuits and systems*, (IEEE, Piscataway, 2020), pp. 213–216
32. Y. Youn et al., Investigation on the worst read scenario of a ReRAM crossbar array. IEEE Trans. Very Large Scale Integr. Syst. **25**(9), 2402–2410 (2017)
33. H. Lim et al., ReRAM crossbar array: Reduction of access time by reducing the parasitic capacitance of the selector device. IEEE Trans. Electron Devices **63**(2), 873–876 (2016)
34. P. Ma et al., High-performance InGaZnO-based ReRAMs. IEEE Trans. Electron Devices **66**(6), 2600–2605 (2019)
35. B. Razavi, The StrongARM latch [a circuit for all seasons]. IEEE Solid-State Circuits Magaz. **7**(2), 12–17 (2015)
36. M.E. Sinangil et al., A 7-nm compute-in-memory SRAM macro supporting multi-bit input, weight and output and achieving 351 TOPS/W and 372.4 GOPS. IEEE J. Solid State Circuits **56**(1), 188–198 (2021)
37. C. Yu, T. Yoo, T. Kim, K. Chai, B. Kim, A 16K current-based 8T SRAM compute-in-memory macro with decoupled read/write and 1-5bit column ADC, in *IEEE custom integrated circuits conference (CICC)*, (IEEE, Piscataway, 2020), pp. 1–4
38. C. Yu, K. Chai, T. Kim, B. Kim, A zero-skipping reconfigurable SRAM in-memory computing macro with binary-searching ADC, in *IEEE 47th European solid-state circuits conference (ESSCIRC)*, (IEEE, Piscataway, 2021), pp. 1–4
39. A. Biswas, A.P. Chandrakasan, CONV-SRAM: An energy-efficient SRAM with in-memory dot-product computation for low-power convolutional neural networks. IEEE J. Solid State Circuits **54**(1), 217–230 (2019)
40. Y.-D. Chih et al., 16.4 an 89TOPS/W and 16.3TOPS/mm2 all-digital SRAM-based full-precision compute-in memory macro in 22nm for machine-learning edge applications, in *2021 IEEE international solid-state circuits conference (ISSCC)*, (IEEE, Piscataway, 2021), pp. 252–254
41. B. Zimmer et al., A 0.32–128 TOPS, scalable multi-chip-module based deep neural network inference accelerator with ground-referenced signaling in 16 nm. IEEE J. Solid State Circuits **55**(4), 920–932 (2020)

Chapter 3
SRAM-Based Processing-in-Memory (PIM)

Hyunjoon Kim, Chengshuo Yu, and Bongjin Kim

3.1 Introduction

SRAM-based PIM gained popularity from its implementation simplicity using active device-only and compatibility with the standard CMOS logic process. Unlike the DRAM macro that is typically placed off-chip, SRAM macro is implemented on-chip to serve as cache memory. Neural Cache [1] and CSRAM [2] took advantage of the conventional SRAM implementation and re-purposed the macro to run compute operations as well as the normal memory storing operation without significant changes in the hardware.

Despite the advantages, SRAM-based PIM using the yield-optimized 6T bitcell provided by the foundry often produces design concerns regarding the data integrity during write/read operations, voltage scaling issues, and precision limitation during the compute operation from its binary storage data. Large capacitive load on the shared bitline in the SRAM array is the source of write/read disturbances, while also limiting the bitline dynamic range. Common approaches to resolve the disturb issues are implementing a hierarchical bitline structure to isolate read and write operations by adding more transistors or providing a custom-designed bitcell. Scalability issue is a common drawback presented in many analog systems. A mixed-signal PIM macro utilizes a set of highly optimized control voltages that are sensitive to process/temperature variation, hence tuning them for scalability would significantly affect the operability of the macro. Finally, the precision limitation is from an inherent design of the SRAM bitcell storing a binary bit.

H. Kim · C. Yu
Nanyang Technological University, Singapore, Singapore
e-mail: KIMH0003@e.ntu.edu.sg; e190026@e.ntu.edu.sg

B. Kim (✉)
University of California Santa Barbara (UCSB), Santa Barbara, CA, USA
e-mail: bongjin@ucsb.edu

© The Author(s), under exclusive license to Springer Nature Switzerland AG 2023
J.-Y. Kim et al. (eds.), *Processing-in-Memory for AI*,
https://doi.org/10.1007/978-3-030-98781-7_3

41

Recent state-of-the-art SRAM-based PIM architectures attempt to address these issues through different bitcell designs [1–12], new architecture [10–24], improved data flow, and optimized compute operation schemes adopted from other computer architectures [12, 22, 25–27]. Many SRAM-based PIM macros exploit bitline operation that relies on computation in analog domain. In general, SRAM-based PIM macro utilizes one of the following three analog computing methods: bitline current discharge accumulation/averaging, charge-domain accumulation, and voltage-based accumulation. Process variation-induced nonlinearity is a common major concern for any of the three design types, and digital domain-computed SRAM-based PIM works attempt to address the particular issue [22, 25].

This chapter introduces the fundamentals in the implementation of SRAM-based PIM macros. Then, various types of SRAM-based PIM cells with their macro circuits and architecture will be described. Finally, the macro implementations with evaluation results are discussed in the following section.

3.2 SRAM-Based PIM Cell Designs

3.2.1 Standard 6T SRAM-Based PIM

A standard 6T SRAM cell, a conventional embedded cache memory cell for storing single-bit data, can process a binary MAC operation in memory. For implementing a MAC operation in the SRAM cell, a binary input is applied to a WL, and a binary weight is stored in the internal storage nodes of the SRAM cell, Q and Qb, as shown in Fig. 3.1a. A DC low signal is applied to the WL to represent a zero input, and a short positive pulse is applied to the WL to represent +1. Note that a pulse is required (instead of a DC high) to accumulate element-wise multiplication results from the SRAM cells sharing the same bitlines. If an SRAM storage node Q stores a "high" (or a "low"), the stored weight is +1 (or −1). As soon as the input (either a DC low or a short positive pulse) is applied to the WL, a binary multiplication in SRAM cell is performed right away based on the input and the stored weight values, and it contributes to the accumulation result (i.e., a voltage difference in BL and BLb). For instance, if the input is 0, the SRAM cell is disabled due to the DC low signal in the WL, and hence, the bitline voltages do not change, as shown in Fig. 3.1b. If the input is +1 or −1, one of the bitlines (BL or BLb) will discharge its capacitance by developing a discharging path between the bitline and the ground, as shown in Fig. 3.1c, d. As a result, the bitline voltage will drop, and the magnitude of the voltage drop is proportional to the pulse width of the input signal. Both BL and BLb are initially precharged to a high voltage level (VDD in Fig. 3.1), and the magnitude of the voltage drop due to a single SRAM cell is ΔV (i.e., a unit voltage drop per SRAM cell).

The element-wise binary multiplication results (i.e., a bitline voltage drop as much as ΔV for +1 and −1 multiplication results) from each bitcells in the

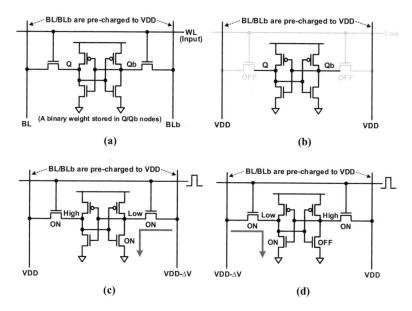

Fig. 3.1 A standard 6T SRAM cell as a PIM cell: (**a**) SRAM cell schematic with input and output for PIM operation; Binary MAC operations when (**b**) the input is zero; (**c**) the input is +1, and the weight is +1; (**d**) the input is +1, and the weight is −1

same column are accumulated and results in an aggregated voltage drop in both bitlines. Figure 3.2, left, illustrates a column of 6T SRAM cells that accumulates element-wise bitline voltage drops when the number of SRAM cells having the multiplication results of +1 (or −1) is P (or N). Based on P and N values, we can estimate the column accumulation result as a bitline voltage difference, $V(BL) - V(BLb) = (P - N) \cdot \Delta V$. Figure 3.2, right, plots the BL (or BLb) voltage as a function of P (or N). Note that a dynamic range is set to ensure a linear accumulate operation and no disturbance issue (i.e., a false SRAM write operation due to a wide bitline dynamic range).

3.2.2 Custom SRAM Cells for PIM

The standard 6T SRAM cell uses bitlines for write and read (or compute for PIM) operations. As a result, there is a disturbance issue that could overwrite SRAM cells with unintended values. Recently, custom SRAM cells with extra transistors have been developed and used for processing MAC operations to prevent the SRAM cell read disturbance issue by decoupling the SRAM write and read ports.

An 8-transistor (8T) foundry SRAM cell with a decoupled read port was used as a PIM unit cell [5]. Two extra NMOS transistors and two additional read ports have been added to decouple the read operation, as shown in Fig. 3.3. The two

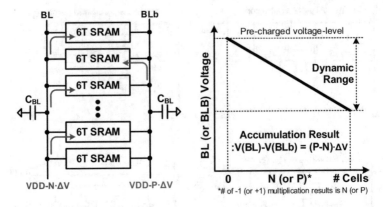

Fig. 3.2 Accumulation in a column of standard 6T SRAM cells

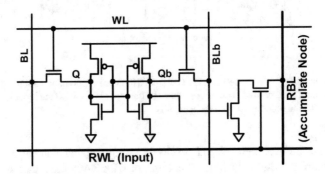

Fig. 3.3 A foundry 8T SRAM cell with a decoupled read port as a PIM unit cell [5]

NMOS transistors are connected in series, and they are used to create a read bitline (RBL) discharging path to the ground. When both an internal SRAM node (Qb) and a read wordline (RWL) input is high, the RBL discharging path is enabled. While the foundry 8T SRAM cell prevents the SRAM read disturbance issue, there are drawbacks, including a single-ended read operation and a larger bitcell area than the 6T standard cell.

A custom 8T SRAM cell has been developed [6] to improve the read operation through differential accumulation nodes. As illustrated in Fig. 3.4, two extra NMOS transistors are added to realize a differential read bitline (RBL and RBLb). Instead of connecting the NMOS transistors to the ground, an input node, RWL, is connected to the source nodes of both the read NMOS transistors. Therefore, a discharging path is created when the RWL is low and the internal circuit node (Q or Qb) is high.

The dynamic range of the bitline discharge is improved from the standard 6T design while the size of the bitcell is increased. Aside from the area increase, short pulse-width control for the RWL is an issue, while the data integrity concern is resolved by the separate read/write ports. The SRAM supply is lowered improving the linearity and reducing the energy consumption, while the precharge voltage for

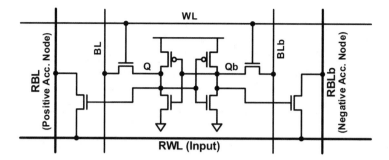

Fig. 3.4 A custom 8T SRAM PIM cell with differential decoupled read ports [6]

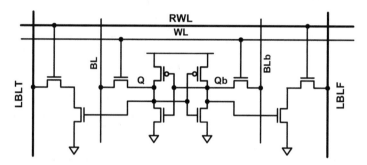

Fig. 3.5 A custom 10T SRAM PIM cell with hierarchical bitlines (LBLT, LBLF) [10]

RWL and RBL/RBLb is set to a higher voltage to guarantee the operation linearity and to resolve leakage concerns.

A custom 10T SRAM cell shown in Fig. 3.5 provides similar improvements to the custom 8T SRAM in terms of the read operation with the differential read and de-coupling read/write ports, however, uses an extra NMOS to load the multiplication result only to the LBLT and LBLF instead of functioning as the accumulation node. 10T implementation further improves the dynamic range of the bitline and also resolves the data integrity issue while sacrificing the bitcell area.

A custom dual 7T SRAM was developed from the custom 8T design to enable reconfigurable weight with 3–15 precision levels in the analog PIM macro, while also de-coupling the write/read operations as shown in Fig. 3.6. The dual 7T can store ternary (3-level) weight values that can form multiple SRAM stacks to represent up to 15-level weight values (3× 7T SRAMs). Also, zero-skipping is implemented for both zero-weight and zero-input for further energy reduction.

In summary, the dual 7T SRAM kept its strengths such as differential read scheme and the improved bitline dynamic range (vs. 6T) with the added features of precision reconfigurability and zero-input/weight skipping, while sacrificing the bitcell area.

The SRAM-based PIM with bitline discharging operations suffers from a limited dynamic range. As the bitline dynamic range increases, the accumulation linearity

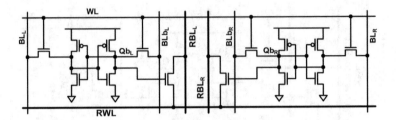

Fig. 3.6 A dual 7T SRAM PIM cell with separate read bitlines (RBL$_L$, RBL$_R$) [11]

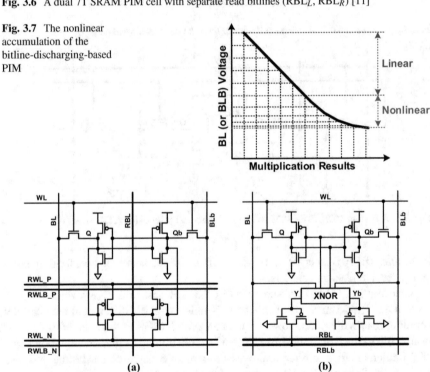

Fig. 3.7 The nonlinear accumulation of the bitline-discharging-based PIM

Fig. 3.8 Custom SRAM-based PIM cells for voltage-mode accumulation: (**a**) single-ended [7] and (**b**) differential accumulation [8]

is significantly degraded, as shown in Fig. 3.7. The voltage-mode accumulation has been introduced to improve the dynamic range in the accumulation nodes [7, 8]. A single-ended voltage-mode accumulation is performed using two CMOS inverters, four RWLs, and a single RBL [7], as shown in Fig. 3.8a. The dynamic range was further improved with a differential voltage-mode accumulation, implemented using two CMOS inverters and two RBLs [8], as shown in Fig. 3.8b. While the voltage-mode PIM macros offer a wide dynamic range and improved linearity, the increased cell size and the residual analog nonidealities are the remaining challenges.

Fig. 3.9 Custom
SRAM-based PIM cells for
charge-domain accumulation:
(**a**) charge-sharing-based; (**b**)
charge-redistribution-based
(or capacitive-coupling-
based)

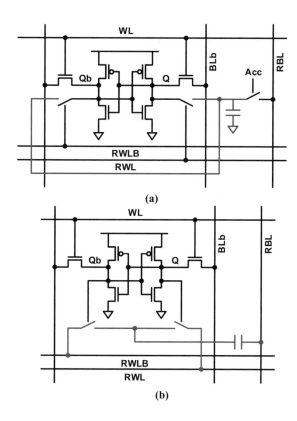

Charge-domain SRAM-based PIM cells [9, 25] have been developed to minimize
the residual analog nonidealities. Instead of using pull-down NMOS transistors (for
bitline discharging) or CMOS (pull-up or pull-down) drivers (for voltage mode
accumulation), the charge-domain SRAM-based PIM cells use passive capacitors
for sharing or redistributing charges. Figure 3.9a shows a PIM cell for charge-
sharing-based accumulation. The cell requires two read wordlines (RWL and
RWLb), a read bitline (RBL), a unit capacitor, and three switches. Figure 3.9b shows
a charge-domain PIM cell with a charge-redistribution-based accumulation using
two read wordlines (RWL and RWLb), a read bitline (RBL), two switches, and a
unit capacitor.

Despite the recent efforts, there are more challenges in the design of analog
and mixed-signal SRAM-based PIM macros, such as data conversion overhead
(i.e., DAC for input conversion and ADC for output conversion), limited reconfigurability, and the residual analog nonidealities (device mismatch, variation, and
nonlinearity). To overcome such limitations, digital SRAM-based PIM cells are
developed [22, 25]. Instead of using analog MAC circuits, the digital PIM cells
use all-digital MAC circuits using an XNOR (or AND) gate and a full-adder.

Hence, the digital PIM is free from analog nonidealities and data conversion
overhead challenges. Figure 3.10a shows a digital PIM cell composed of a standard

Fig. 3.10 Digital
SRAM-based PIM cells: (**a**) a
fixed weight precision PIM
cell [25]; (**b**) a reconfigurable
weight precision PIM cell
[22]

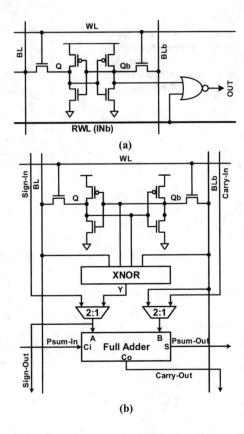

(a)

(b)

6T SRAM cell and a NOR gate working as an AND gate for a bitwise multiplier.
The accumulation is performed in a digital adder tree at the macro-level. Figure
3.10b shows a reconfigurable digital PIM cell that embeds a bitwise XNOR-gate
multiplier and a full-adder for accumulation. Note that a digital PIM macro based
on the reconfigurable PIM cell can be reconfigured to operate with a 1–16b variable
weight precision. A bit-serial computing scheme is applied for inputs for both digital
PIM cells, saving a significant area for conventional bit-parallel computing. The
bit-serial computing is also intrinsically reconfigurable as the number of bit-serial
operation cycles determines the serialized input bit precision.

3.3   SRAM-Based PIM Macro Designs

Figure 3.11 shows four columns of foundry 8T SRAM cells for processing a dot-
product computation between 4b inputs and 4b weights stored in the SRAM cells. A
4b weight is stored bit-by-bit into four SRAM cells in the same row. A 4b serialized
input is applied to a horizontal RWL as multiple positive short pulses. The multiple

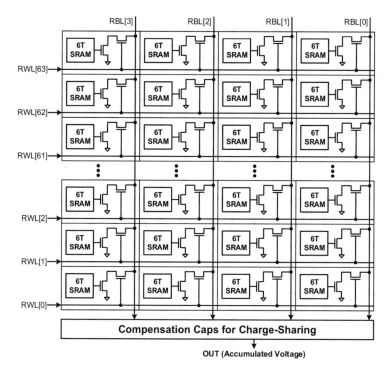

Fig. 3.11 Four columns of foundry 8T SRAM cells for a dot-product PIM operation between 4b inputs and 4b weights [5]

pulse scheme can guarantee better linearity than the pulse-width modulation (PWM) scheme. Each column first accumulates element-wise multiplication results based on the bitline discharging method. Then all four column accumulation results are combined by charge-sharing across the bitlines from the four columns.

The detailed bitline-discharging and charge-sharing operations using four SRAM-based PIM columns are illustrated in Fig. 3.12. It consists of four-step operations: (1) RBL precharging; (2) RBL discharging (column accumulations); (3) charge-sharing; (4) analog-to-digital conversion of the combined accumulation results. RBL [3:0] is the first precharged to a high level (step 1) and then is discharged based on the multiplication results between a series of multiple input pulses (RWL [63:0]) and the weights stored in each column (step 2). After the column-by-column accumulation based on RBL discharging, charge-sharing (step 3) combines the accumulated results. Finally, the combined analog voltage is converted to a digital output code (step 4).

Figure 3.13 shows PIM macro architecture using the foundry 8T SRAM with column-wise multiply-and-average (MAV) scheme using 64× 4b inputs and 16× 4b weights in a single cycle to produce 16× 4b outputs. Each four RBLs (i.e., 4× SRAMs in the same row to realize 4b weight) share one 4b flash ADC, while the RWL counters produce pulse control signals to realize input precision of 4b. RWL

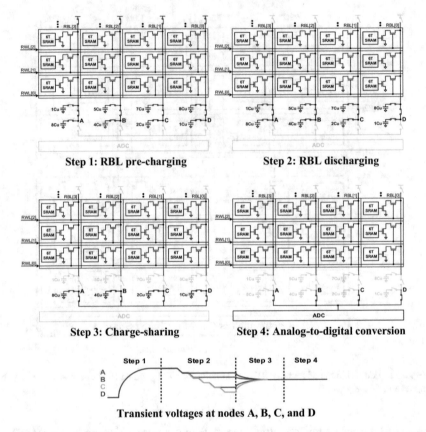

Fig. 3.12 Four-step operations for bitline-discharging and charge-sharing operations to combine accumulation results from four columns [5]. Step 1 RBL precharging. Step 2: RBL discharging. Step 3: Charge-sharing. Step 4: Analog-to-digital conversion. Transient voltages at nodes A, B, C, and D

voltage is sampled through the compensation cap, then the averaging is processed by charge sharing from the binary-weighted computation caps (included in the flash ADC), while the inherent capacitance of the sense amplifier inside 4b flash ADCs used to represent the unit capacitance that effectively produces MAV results (averaged voltage) at the output. The architecture achieves high energy efficiency using advanced technology and novel computation blocks; however, the design could not completely resolve the limited dynamic range of the bitline as well as the process variation induced nonlinearity.

Figure 3.14 illustrates a column-based circuit for processing a dot-product using custom 8T SRAM PIM cells (Fig. 3.6) [6] with differential read ports for bitline-discharging-based accumulation. A column consists of a precharge PMOS circuit, 128× SRAM PIM cells (64× for dot-product, 32× for ADC, and 32× for offset calibration) and a sense amplifier (SA) for a single-slope column ADC operation.

Fig. 3.13 A PIM macro architecture of a 64 × 64 foundry 8T SRAM cell array with 64× 4b inputs and 16× 4b outputs [5]

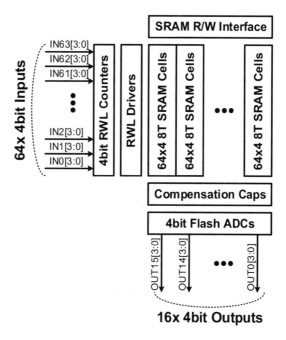

Figure 3.14, right, plots transient waveforms for maximum and minimum RBL voltages based on the number of cells discharging RBL or RBLb. The bitline capacitances are discharged for a short RWL negative pulse width.

In this particular design, the custom 8T SRAM offered reduced ADC overhead by using the replica bitcells, improved the bitline dynamic range compared to the standard 6T SRAM and addressed variation-induced nonlinearity through offset calibration blocks. On the other hand, the dynamic range improvement is not significant, ADC overhead is not completely resolved, and the parallel MAC utilization is reduced in half due to the bitcells that are assigned as column ADCs and offset calibration blocks.

A PIM macro architecture using the custom 8T SRAM is shown in Fig. 3.15, and its column MAC structure is illustrated in Fig. 3.14. 128× input rows are divided into three functional blocks for inputs, ADC reference, and offset calibration to perform column-wise MAC operation to output 128× 1b outputs. The custom 8T SRAM macro utilizes wordline pulse width modulation and reconfigurable ADC precision through replica PIM cells to avoid larger ADC overhead while sacrificing the input parallelism. Compared to the architecture in Fig. 3.13, the custom 8T SRAM macro provides higher area efficiency from the non-flash ADC implementation and higher number of input columns.

Single-ended voltage-mode SRAM PIM cells [7] are used for processing a dot-product in a column-based memory array, as shown in Fig. 3.16. XNOR-based multiplications are performed between the stored binary weights (in a 6T standard SRAM cell embedded in each PIM cell) and the input read wordlines connected

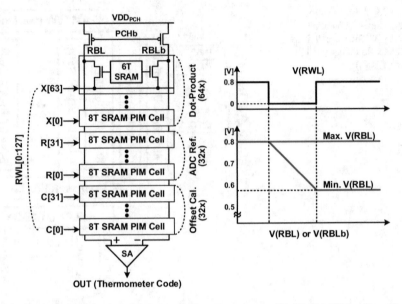

Fig. 3.14 A column-based dot-product circuit using custom 8T SRAM PIM cells with an embedded single-slope column ADC based on replica bitcells [6]

Fig. 3.15 A PIM macro architecture of a 128 × 128 custom 8T SRAM cell array [6]

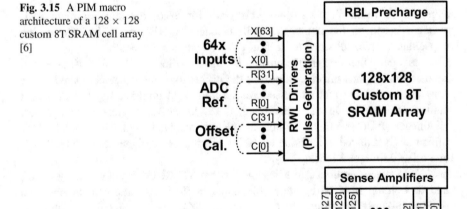

to the supplies and grounds of voltage-mode drivers. As a result, a shared RBL is driven by parallel pull-up and pull-down transistors equivalent to a resistive divider, as illustrated in Fig. 3.16, top-right. The measured transfer characteristic of the single-ended voltage-mode accumulator is shown in Fig. 3.16, bottom-right. Note that the RBL dynamic range is rail-to-rail, which has been significantly

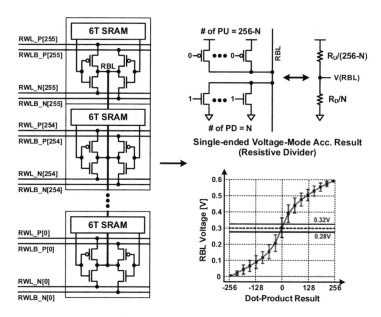

Fig. 3.16 A column-based dot-product circuit using single-ended voltage-mode SRAM PIM cells in Fig. 3.8a and the measured accumulated RBL voltage [7]

improved from that of the bitline-discharging-based current-mode accumulations using standard 6T or foundry/custom 8T PIM cells [1, 2]. However, there are residual analog nonidealities (nonlinearity and variation), as is shown in Fig. 3.16, bottom-right.

The architecture of the single-ended voltage-mode SRAM PIM macro is presented in Fig. 3.17. 256× ternary input is processed through column-wise MAC. XNOR-based multiplication of the PIM macro architecture utilizes voltage accumulation which requires separate computing blocks that increase the size of the bitcell. Also, the MAC results are selectively fed to the multi-bit flash ADC due to area overhead to produce 10b thermometer code. While the voltage-mode operation provides improved dynamic range in the RBL, the reduction in throughput due to A2D conversion overhead, analog variation-induced nonlinearities and the nonlinear accumulation caused by the CMOS pull-up/pull-down strength imbalance are major concerns.

Differential voltage-mode SRAM PIM cells [8] are developed and used for processing a dot-product, as shown in Fig. 3.18. Embedded XNOR gates are used for multiplications between the stored binary weights in the standard 6T SRAM cells and the inputs applied through SRAM bitlines. Note that the SRAM bitlines are reused for the PIM operations to minimize the overhead due to the extra input signal lines. After the element-wise multiplications, differential voltage-mode drivers (i.e., a pair of inverters) are used for voltage-mode accumulation. Each bitline (RBL and RBLb) is pulled up and pulled down based on the number of $+1$ and -1

Fig. 3.17 A PIM macro
architecture of a 256 × 64
single-ended voltage-mode
SRAM cell array with 256×
ternary inputs and a 10b
thermometer code output [7]

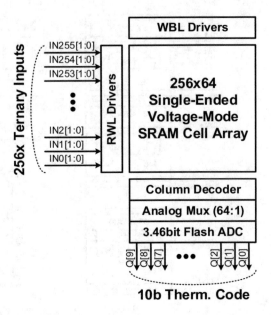

multiplication results. Figure 3.18, bottom-left, shows the equivalent circuits representing the pull-up and pull-down resistors that eventually determine the voltages for RBL and RBLb. The Monte-Carlo simulation result of the pseudo-differential RBL voltage (i.e., V(RBL)-V(RBLb)) is shown in Fig. 3.18, bottom-right. Note that a dynamic range is doubled and the transfer characteristic is symmetric while residual nonlinearities and variations are observed.

Differential voltage-mode SRAM PIM macro architecture is shown in Fig. 3.19. Similar to the custom 8T SRAM PIM, 128× inputs are divided into the same three functional parts in row-wise compute scheme. Although the operational difference from the custom 8T SRAM PIM exists due to the voltage mode operation as opposed to the bitline discharge, a similar performance trade-off of reduced parallelism exists from the column bitcells that are assigned for row ADC and offset calibration. Also, the low precision weight SRAMs limit the application mapping as well as the PVT variation-induced nonlinearity causing output reliability issues.

Compared to the architecture described in Fig. 3.17, differential voltage-mode SRAM PIM provides higher throughput due to the elimination of the output analog MUX and minimized ADC overhead from the replica bitcells that realize parallel row ADCs.

Passive capacitors have been used to implement charge-domain SRAM-based PIM macros to minimize analog nonidealities in processing accumulations using bitline-discharging or voltage-mode accumulation. Figure 3.20 shows a column of custom SRAM PIM cells for charge-sharing-based accumulation. A PIM cell consists of a standard 6T SRAM cell, a pair of switches, and two read wordline inputs (RWL and RWLb) for an XNOR-based binary multiplication. A unit passive capacitor and a switch are used for sharing charges across the bitcells in the same

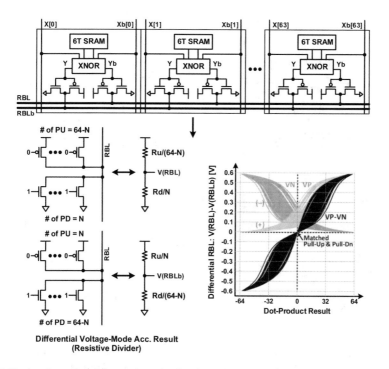

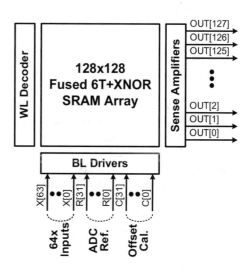

Fig. 3.18 A column-based dot-product circuit using differential voltage-mode SRAM PIM cells in Fig. 3.8b and the Monte-Carlo simulated differential RBL voltage [8]

Fig. 3.19 A PIM macro with a 128 × 128 differential voltage-mode SRAM PIM cells [8]

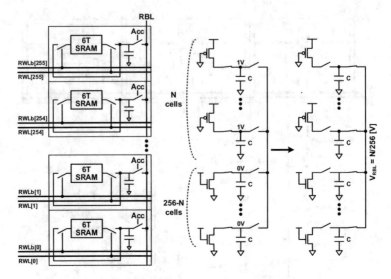

Fig. 3.20 A column-based dot-product circuit using SRAM-based PIM cells in Fig. 3.9a for accumulation based on charge-sharing and the resulting RBL voltage

column. Figure 3.20, right, illustrates the charge-sharing-based accumulation before and after turning on the accumulator switches all at once. When the number of +1 element-wise multiplication results is "N," the resulting RBL voltage is $N/256$ V if the supply voltage is 1 V.

Besides the bitline charge-sharing, a charge-redistribution (or capacitive coupling) technique can be used for implementing a charge-domain SRAM-based PIM macro. Figure 3.21 shows a column of SRAM-based PIM cells realizing charge-redistribution-based accumulation. Each PIM cell consists of a standard 6T SRAM cell, a pair of switches with two input read wordlines (RWL and RWLb), and a unit capacitor.

Charge domain SRAM-based PIM macros offer the highest energy efficiency and throughput performances among the other state-of-the-art mixed-signal SRAM-based PIM works. A tradeoff for such performance for the work described in Fig. 3.21, however, are the nonlinearity issue at the output caused by the shared capacitive coupling of the ADC and the limited DNN applicability due to low precision parameter storage of the macro.

Charge-redistribution-based SRAM PIM macro architecture is illustrated in Fig. 3.22. 256× inputs are processed through column-wise MAC producing 64× 4b ADC outputs. The macro can implement binary neural networks (BNN) and achieves high throughput by enabling all the RWL rows simultaneously instead of using the conventional row-by-row or column-by-column access. An analog voltage is formed at the read bitline through capacitive division, then fed to the ADC placed at the output of each column, and produces fully parallel MAC results in a single cycle.

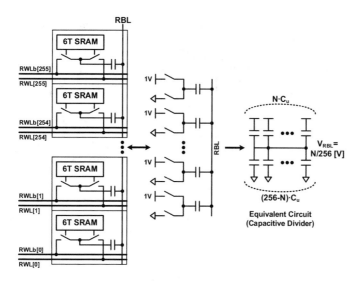

Fig. 3.21 A column-based dot-product circuit using SRAM-based PIM cells in Fig. 3.9b for accumulation based on charge-redistribution (or capacitive coupling)

Fig. 3.22 A PIM macro of 256 × 64 charge-redistribution-based SRAM PIM cells [9]

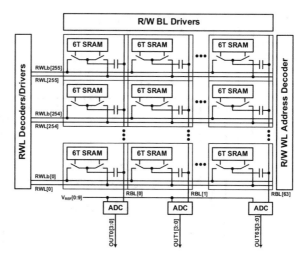

10T SRAM-based row-wise PIM structure is shown in Fig. 3.23a. Hierarchical bitlines (LBLT/F and BL/B) provide robustness toward the write disturbance caused by the large capacitive load during the current discharge. Multiply-and-average results are applied to the V_p and V_n nodes then integrated through charge-sharing ADC reference column as shown in Fig. 3.23b. Each 10T SRAM stores 1b filter weights for Convolutional Neural Network (CNN) inference, while the integrated MAV produces 7b ADC output, which is described in Fig. 3.24. The macro architecture shows row-wise computing scheme producing 16× 7b parallel outputs with 7b input DAC that utilizes pulse-width modulated wordline control. Despite

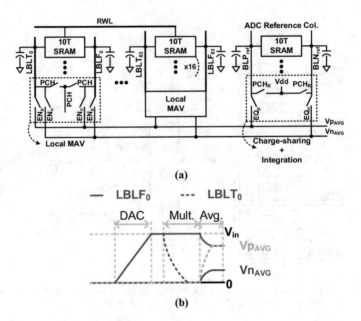

(a)

(b)

Fig. 3.23 10T SRAM-based PIM architecture: (**a**) a row-wise charge sharing architecture; (**b**) its multiply-and-average (MAV) operation diagram [10]

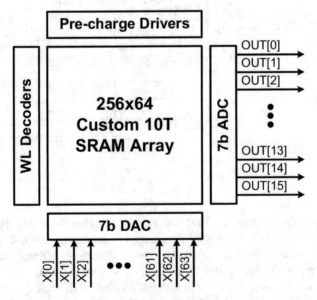

Fig. 3.24 A PIM macro architecture of a 256 × 64 10T SRAM cell array with 64× 7b inputs and 16× 7b outputs [10]

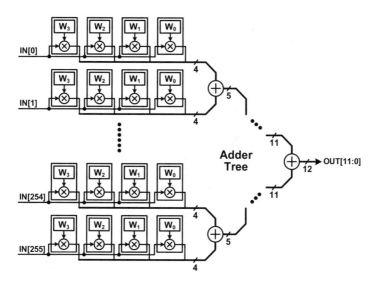

Fig. 3.25 A digital SRAM PIM dot-product circuit using 256 × 4 PIM cells and an adder tree [25]

the larger SRAM area, the area efficiency did not suffer due to the small size of the CSH_ADC. However, hardware scalability, low precision weight operation and analog variation issues remain as design challenges.

Analog PIM works [1–4, 13, 16] achieve outstanding energy efficiency and energy efficiency. However, analog computing issues such as ADC/DAC overhead and PVT variation-induced nonlinearities remain as major concerns. Digital PIM architectures address both issues by directly processing the digital input bits without the data conversion and utilizing the binary abstraction that reduces sensitivity to any physical variation.

Figures 3.25 and 3.26 illustrate digital PIM macro [25] that utilize 256× 4b weights per column, producing 64× 12b MAC output. Each PIM cell is composed of fused 6T SRAM and a two-input NOR gate producing binary multiplication result while the accumulate operation is processed separately in the dedicated adder tree that uses 256× 4b multiply results as inputs to produce a single 12b output. Although the weight bit precision is fixed to 4b, bit-width can be further reconfigured to 8b, 12b, and 16b depending on the multiple macro scaling with area trade-off.

Figures 3.27 and 3.28 present another digital PIM macro [22] with fully reconfigurable inputs, weights, and outputs. Each PIM cell is composed of 6T SRAM, an XNOR gate and a full-adder. The PIM cells can form 1–16b unit column MAC by stacking together to achieve reconfigurable processing element (PE). Despite the regular design of unit PIM cells in a multi-bit column MAC, different function is assigned depending on its location within the unit column.

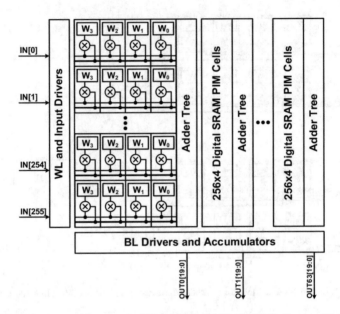

Fig. 3.26 A PIM macro architecture of a 256 × 256 digital SRAM PIM cell array with adder trees [25]

Fig. 3.27 A reconfigurable digital SRAM PIM dot-product circuit using (N+7) × 128 PIM cells [22]

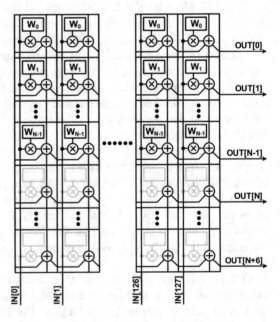

For example, the PIM cell located at the top of the column represents LSB weight and the gray cells shown in Fig. 3.26 are accumulation-only PIM cells. The gray PIM cells produce sign-extended output as well as partial sum propagation through

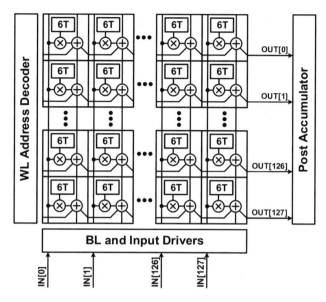

Fig. 3.28 A PIM macro of 128×128 reconfigurable digital SRAM PIM cells [22]

all the columns in the macro array. With $128\times$ column, each unit column MAC requires $7\times$ accumulation-only PIM cells to propagate all the partial sums to the output. Challenges associated with the digital SRAM PIM [22] is the hardware redundancy caused by enabled/disabled memory and compute blocks from the regular structure that processes reconfigurable MAC operation. As a result of the redundant circuit blocks, unit PIM cell becomes large and the SRAM capacity in the macro is degraded.

3.4 Summary

This chapter provides a review of recent state-of-the-art SRAM-based PIM macros. SRAM-based PIMs are currently the most popular implementations due to their manufacturing compatibility with the standard CMOS logic process. Unlike DRAM or other emerging memory technologies, standard yield optimized SRAM bitcell does not require a special manufacturing process. As a result, SRAM-based PIM could explore diverse custom bitcell designs without a major concern in manufacturability. Most state-of-the-art SRAM-based PIM targets optimizing the high-level architecture and data-flow rather than focusing on reliability and stability of the physical implementation of the macro. Hence, SRAM-based PIM architectures provide a primitive basis of the advanced architecture implementation for the more process-sensitive DRAM-based PIM and ReRAM-based PIM macros in terms of operation techniques and implementation.

SRAM-based PIM macro can be categorized to two different types based on its operation principles. Analog PIM macro provides high energy/area efficiency performance while showing limitations in flexibility and reliability. Different SRAM designs were proposed to address critical issues associated with the analog computing and memory operations. On the other hand, digital PIM macro demonstrates less efficient numbers but excels in reconfigurability and robustness to variations. Digital SRAM-based PIM works prioritize the standard 6T SRAM design, as the most critical design challenge for a digital macro is reducing the hardware footprint and energy consumption. The two design approaches supplement each other's weaknesses and a recent publication in PIM research introduced hybrid architecture [12] that combines the strengths of the two types.

References

1. C. Eckert et al., Neural cache: Bit-serial in-cache acceleration of deep neural networks, in *ACM/IEEE 45th annual international symposium on computer architecture (ISCA)*, (ACM, New York, 2018), pp. 383–396
2. J. Wang et al., 14.2 a compute SRAM with bit-serial integer/floating-point operations for programmable in-memory vector acceleration, in *2019 IEEE international solid-state circuits conference - (ISSCC)*, (IEEE, Piscataway, 2019), pp. 224–226
3. X. Si et al., 24.5 a twin-8T SRAM computation-in-memory macro for multiple-bit CNN-based machine learning, in *2019 IEEE international solid-state circuits conference - (ISSCC)*, (IEEE, Piscataway, 2019), pp. 396–398
4. Q. Dong, S. Jeloka, Y. Kim, M. Kawaminami, A. Harada, S. Miyoshi, M. Yasuda, D. Blaauw, D. Sylvester, A 4 + 2T SRAM for searching and in-memory computing with 0.3-V V_{DDmin}. IEEE J. Solid State Circuits **53**(4), 1006–1015 (2018)
5. Q. Dong, M.E. Sinangil, B. Erbagci, D. Sun, W. Khwa, H. Liao, Y. Wang, J. Chang, A 351TOPS/W and 372.4GOPS compute-in-memory SRAM macro in 7nm FinFET CMOS for machine-learning applications, in *IEEE int. solid-state circuits conf. (ISSCC)*, (IEEE, Piscataway, 2020), pp. 242–244
6. C. Yu, T. Yoo, T. Kim, K. Chai, B. Kim, A 16K current-based 8T SRAM compute-in-memory macro with decoupled read/write and 1- 5bit column ADC, in *IEEE custom integrated circuits conference (CICC)*, (IEEE, Piscataway, 2020), pp. 1–4
7. S. Yin, Z. Jiang, J.-S. Seo, M. Seok, XNOR-SRAM: In-memory computing SRAM macro for binary/ternary deep neural networks. IEEE J. Solid State Circuits **55**(6), 1733–1743 (2020)
8. H. Kim, Q. Chen, B. Kim, A 16K SRAM-based mixed-signal in-memory computing macro featuring voltage-mode accumulator and row-by-row ADC, in *IEEE Asian solid-state circuit conference (ASSCC)*, (IEEE, Piscataway, 2019), pp. 35–36
9. Z. Jiang, S. Yin, J. Seo, M. Seok, C3SRAM: An in-memory-computing SRAM macro based on robust capacitive coupling computing mechanism. IEEE J. Solid State Circuits **55**(7), 1888–1897 (2020)
10. A. Biswas, A.P. Chandrakasan, CONV-SRAM: An energy-efficient SRAM with in-memory dot-product computation for low-power convolutional neural networks. IEEE J. Solid State Circuits **54**(1), 217–230 (2019)
11. C. Yu, K. Chai, T. Kim, B. Kim, A zero-skipping reconfigurable SRAM in-memory computing macro with binary-searching ADC, in *IEEE European solid-state circuit conference (ESSCIRC)*, (IEEE, Piscataway, 2021), pp. 131–134

12. J. Kim, J. Lee, J. Heo, J.Y. Kim, Z-PIM: A sparsity-aware processing-in-memory architecture with fully variable weight bit-precision for energy-efficient deep neural networks. IEEE J. Solid State Circuits **56**(4), 1093–1104 (2021)
13. W. Khwa et al., 31.5 a 65nm 4Kb algorithm-dependent computing-in-memory SRAM unit-macro with 2.3ns and 55.8TOPS/W fully parallel product-sum operation for binary DNN edge processors, in *2018 IEEE international solid-state circuits conference - (ISSCC)*, (IEEE, Piscataway, 2018), pp. 496–498
14. X. Si et al., 15.5 a 28nm 64Kb 6T SRAM computing-in-memory macro with 8b MAC operation for AI edge chips, in *2020 IEEE international solid-state circuits conference - (ISSCC)*, (IEEE, Piscataway, 2020), pp. 246–248
15. J. Su et al., 15.2 a 28nm 64Kb inference-training two-way transpose multibit 6T SRAM compute-in-memory macro for AI edge chips, in *2020 IEEE international solid-state circuits conference - (ISSCC)*, (IEEE, Piscataway, 2020), pp. 240–242
16. H. Jia, M. Ozatay, Y. Tang, H. Valavi, R. Pathak, J. Lee, N. Verma, 15.1 a programmable neural-network inference accelerator based on scalable in-memory computing, in *2021 IEEE international solid-state circuits conference - (ISSCC)*, (IEEE, Piscataway, 2021), pp. 236–238
17. J. Yue et al., 15.2 a 2.75-to-75.9TOPS/W computing-in-memory NN processor supporting set-associate block-wise zero skipping and ping-pong CIM with simultaneous computation and weight updating, in *2021 IEEE international solid-state circuits conference - (ISSCC)*, (IEEE, Piscataway, 2021), pp. 238–240
18. R. Guo et al., 15.4 a 5.99-to691.1 TOPS/W tensor-train in-memory-computing processor using bit-level-sparsity-based optimization and variable-precision quantization, in *2021 IEEE international solid-state circuits conference - (ISSCC)*, (2021), pp. 242–244
19. J. Su et al., 16.3 a 28nm 384kb 6T-SRAM computation-in-memory macro with 8b precision for AI edge chips, in *2021 IEEE international solid-state circuits conference - (ISSCC)*, (IEEE, Piscataway, 2021), pp. 250–252
20. M. Kang, S. Gonugondla, A. Patil, N. Shanbhag, A multi-functional in-memory inference processor using a standard 6T SRAM array. IEEE J. Solid State Circuits **53**(2), 642–655 (2018)
21. J. Zhang, Z. Wang, N. Verma, In-memory computation of a machine-learning classifier in a standard 6T SRAM array. IEEE J. Solid State Circuits **52**(4), 915–924 (2017)
22. H. Kim, T. Yoo, T. Kim, B. Kim, Colonnade: A reconfigurable SRAM-based digital bit-serial compute-in-memory macro for processing neural networks. IEEE J. Solid State Circuits **56**(7), 2221–2233 (2021)
23. H. Valavi, P. Ramadge, E. Nestler, N. Verma, A 64-tile 2.4-Mb in-memory-computing CNN accelerator employing charge-domain compute. IEEE J. Solid State Circuits **54**(6), 1789–1799 (2019)
24. K. Ando et al., BRein memory: A single-chip binary/ternary reconfigurable in-memory deep neural network accelerator achieving 1.4 TOPS at 0.6 W. IEEE J. Solid State Circuits **53**(4), 983–994 (2018)
25. Y. Chih et al., 16.4 an 89TOPS/W and 16.3TOPS/mm^2 all-digital SRAM-based full-precision compute-in memory macro in 22nm for machine-learning edge applications, in *2021 IEEE international solid- state circuits conference (ISSCC)*, (IEEE, Piscataway, 2021), pp. 252–254
26. H. Sharma et al., Bit fusion: Bit-level dynamically composable architecture for accelerating deep neural network, in *Proc. 45th annual IEEE/ACM international symposium on computer architecture (ISCA)*, (ACM, New York, 2018), pp. 764–775
27. P. Judd et al., Stripes: Bit-serial deep neural network computing, in *Proc. 49th Annual IEEE/ACM international symposium on microarchitecture (MICRO)*, (ACM, New York, 2016), pp. 1–12

Chapter 4
DRAM-Based Processing-in-Memory

Donghyuk Kim and Joo-Young Kim

4.1 Introduction

This chapter explains various PIM architectures and implementations based on
DRAM technology. As illustrated in Fig. 4.1, we triage the DRAM-based PIMs
into three categories based on the level of logic integration. The first category is
the low-level PIM, which integrates logic with bitline sense amplifiers to utilize
the memory bank's internal bandwidth fully. AMBIT [1] and DRISA [2] are
representative works. Second, Newton [3] by SK Hynix integrates multiply-and-
accumulate (MAC) units at the bank level, after the data sense amplifiers. Although
this method cannot utilize the maximum internal bandwidth as it integrates logic
after the column decoder, it has more space for logic integration. HBM-PIM [4]
by Samsung Electronics integrates processing units at the same level but for high-
bandwidth memory (HBM), a 3-d stacked memory. The last category is the 3-D
PIM, which utilizes the whole vertical stack of a 3-d stacked memory including the
base logic die for in-memory processing. Neurocube [5], Tetris [6], and iPIM [7]
fall into this category.

4.2 Basic DRAM Operation

Figure 4.2 shows the typical organization of a modern DRAM chip. It broadly
consists of control logic, multiple banks, and data IO circuitry. The DRAM bank is

The original version of the chapter has been revised. A correction to this chapter can be found at
https://doi.org/10.1007/978-3-030-98781-7_9.

D. Kim · J.-Y. Kim (✉)
School of Electrical Engineering (E3-2), KAIST, Daejeon, South Korea
e-mail: kar02040@kaist.ac.kr; jooyoung1203@kaist.ac.kr

Fig. 4.1 DRAM PIM
architectures

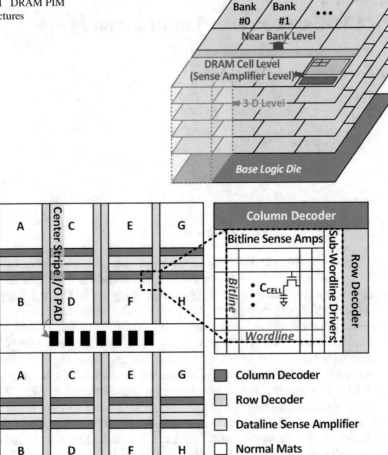

Fig. 4.2 DRAM chip organization

made of DRAM mat, the basic 2-dimensional array structure with periphery circuits
to access the cells. In the mat, the row decoder specifies a single wordline and
drives it based on the address. Then, all the access transistors of the DRAM cells
connected to the wordline are activated, and the values are loaded into the bitlines.
Each DRAM cell, composed of a single transistor and a capacitor, starts to share the
stored charge with the bitline, which is pre-charged to half V_{DD} when the transistor
is turned on. Slight voltage differences in the bitlines caused by the charge sharing
are amplified by the bitline sense amplifiers in the bottom. It is also called row buffer
because it stores the cell values of the entire row. Once an entire row is buffered in
the row buffer, the column decoder chooses one or more bitlines to transfer data to
the IO pads.

4.3   Bulk Bitwise Processing-in-Memory

In this section, we study the DRAM-based PIMs that embed logic into the level of
memory cells/ bitline sense amplifiers, which is the lowest level you can possibly
design. In order to maximally use the internal read bandwidth, they enable multiple
rows at once and perform low-level logic operations such as AND, OR, or NOR for
entire rows at the bitline sense amplifiers. It is called bulk bitwise processing.

4.3.1   AMBIT

4.3.1.1   Triple Row Activation

As shown in Fig. 4.3, AMBIT activates three wordlines simultaneously, unlike the
regular DRAM only activates a single row at a time. If we look from the perspective
of a single bitline, three cells connected to the bitline are accessed at the same cycle.
Then, all the charges from the cells will be shared at the bitline. If the number of cells
charged with V_{DD} (i.e., logical 1) is larger than the number of cells with no charge
(i.e., logical 0), the amount of net charge injecting to the bitline will be positive.
Since the bitline is already pre-charged to half-V_{DD}, the final voltage level by the
charge sharing will be a bit higher than the half-V_{DD} and eventually goes to V_{DD} by
the sense amplifier. In other words, if the bitline has two or three charged cells, the
final voltage will be V_{DD}. On the other hand, if the bitline has zero or one charged
cell, the final voltage value will be 0. Equation 4.1 shows the exact charge sharing
equation among the three cells by the triple row activation (TRA). C_c and C_b are
cell capacity and bitline capacity, respectively, and k is the number of 1s among the
three cells.

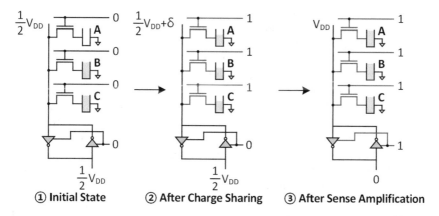

Fig. 4.3 Triple row activation

$$\delta = \frac{k \cdot C_c \cdot V_{DD} + C_b \cdot \frac{1}{2} V_{DD}}{3C_c + C_b} - \frac{1}{2} V_{DD} = \frac{(2k-3)C_c}{6C_c + 2C_b} V_{DD} \qquad (4.1)$$

The final value of the TRA becomes 1 if the number of 1s among three cells is more than or equivalent to 2. It is the same as the majority function. Among the three cell values (i.e., A, B, and C), if either A and B or B and C or C and A are 1, then the final value will be 1. This is $AB + BC + CA$ in a simple Boolean equation, which is same as $C(A + B) + \overline{C}(AB)$. Based on this Boolean equation, we can easily implement AND or OR function by controlling the C value. If C is set to 0, the first term will be gone, so the final value will be AB. This is the AND operation for all the bits between the entire two rows. Otherwise, if C is set to 1, the final value will be $A + B$, the OR operation for the entire two rows. By presetting C and enabling three rows simultaneously, we can implement AND or OR operation for the entire bits of the two rows. As an entire row of a DRAM bank can be multiple kilobytes (usually 1KB or 2KB), this TRA scheme enables multi-kilobytes bitwise AND/OR operation. This is the main idea of the AMBIT. AMBIT utilizes the TRA and the property of charge sharing in the bitlines, as it is difficult to integrate an AND or OR gate within a pitch of the tiny DRAM cell because each gate requires six transistors. Thanks to TRA, AMBIT implements bulk bitwise operations without any transistors added to the sense amplifiers.

4.3.1.2   AMBIT DRAM Organization

Although AMBIT efficiently implements the bulk bitwise AND/OR operation using TRA, it has some issues. First, the TRA re-writes the final result to the original cells like a normal memory read does. As it activates three rows at the same time and gets AND/OR values, it destroys the original values of the cells in the three rows. The second issue is the cost of TRA. Since it needs to activate three rows simultaneously, the decoding logic needs to decode three addresses at once. Because it causes a linear increase in address bus and row decoding logic, the TRA puts a burden on the control logic of the memory cell array.

To solve the above issues, AMBIT divides the row address space of the memory subarray into three groups: (1) bitwise group (B-group), (2) control group (C-group), and (3) data group (D-group), as shown in Fig. 4.4. B-group has eight designated rows for bulk bitwise AND/OR operations with special decoding logic for TRA. C-group pre-stores 0 and 1 to select AND and OR operation, respectively. D-group stores the original data, occupying the most rows in the subarray. For C-group and D-group, AMBIT uses the regular row decoder that does not require any changes in design. For the operation, it copies the two rows of data from D-group to the designated rows in B-group (i.e., T0 and T1). These two rows will be the input operands of the bulk bitwise operation. It also initializes the designated row T2 to 0 (=AND) or 1 (=OR) to choose the operation. Then, it simultaneously activates the three designated rows, T0, T1, and T2, for computation. Finally, it copies the result

Fig. 4.4 AMBIT memory organization

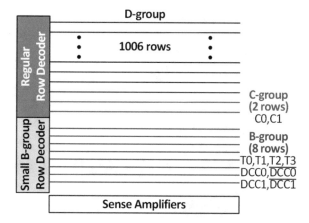

row T0 to a row in the D-group. Using three copies between the main D-group and special-purpose B-group, AMBIT completes the bulk bitwise operation.

4.3.1.3   Fast Row Copy

AMBIT requires a lot of row-wide copies between the main D-group and the special-purpose B-group designated for TRA. To reduce the long latency of the row copies between the two groups, the authors utilize the method of RowClone-FPM (Fast Parallel Mode) [8]. For a row-wide copy, a regular DRAM requires an activation command followed by many column-read commands and the final pre-charge command to read an entire row. It also needs an activation command followed by many column-write commands and the final pre-charge command to write a destination row. Unlike the regular DRAM requires lots of commands with a very long latency of more than 1000 ns, the RowClone-FPM uses only three commands: source row activation, destination row activation, and the pre-charge, as shown in Fig. 4.5. By activating the destination row right after the source row being amplified at the row buffer, RowClone-FPM copies the entire source row to the destination row very efficiently, reducing the latency more than 10 times.

4.3.1.4   Bulk Bitwise NOT

Another problem with AMBIT's AND/OR-based processing is that it is not functionally complete. To be functionally complete, it needs NOT operation. To address this, ABMIT utilizes the inverted value of the bitline sense amplifier. To this end, it introduces a row of dual-contact cells (DCCs), in which each DCC has an additional pair of wordlines and access transistor to move the inverted value of the sense amplifier to the cell. Figure 4.6 illustrates how AMBIT works bulk bitwise NOT operation with DCC. First, it activates the wordline of the source row, and the

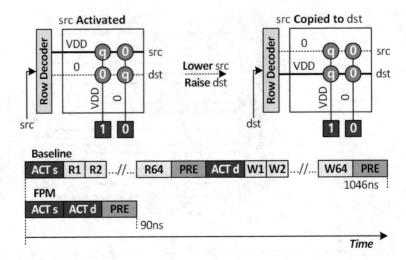

Fig. 4.5 Fast row copy using RowClone-FPM

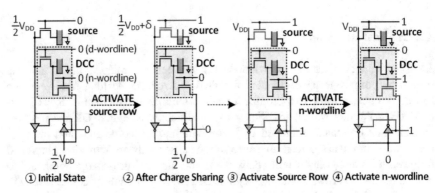

Fig. 4.6 Bulk bitwise NOT with dual-contact cell

sense amplifier evaluates the cell value. Then, it enables the n-wordline of the DCC, which connects the inverted node of the sense amplifier and the cell in the DCC. If the inverted value of the bitline is 0, the charge in the cell will be discharged to the ground. Hence, the cell value of DCC is the same as the inverted value. On the other hand, if the inverted value is 1, the cell will be charged to V_{DD}. Therefore, AMBIT successfully moves the inverted value of the source row to the DCC using the n-wordline and its transistor. DCC uses d-wordline and its transistor when it needs to copy the row of DCCs, the result of NOT operation, to another row.

The actual implementation of DCC may not be feasible as it requires one more wordline and transistor to fit the pitch of a DRAM cell. In DRAM, the pitch of a single cell is already optimized to having only a single transistor and a capacitor. Adding another transistor and a wordline to it can be extremely challenging (Table 4.1).

Table 4.1 B-group address table

Addr.	Wordline(s)	Addr.	Wordline(s)
B0	$T0$	B8	$\overline{DCC0}, T0$
B1	$T1$	B9	$\overline{DCC1}, T1$
B2	$T2$	B10	$T2, T3$
B3	$T3$	B11	$T0, T3$
B4	$DCC0$	B12	$T0, T1, T2$
B5	$\overline{DCC0}$	B13	$T1, T2, T3$
B6	$DCC1$	B14	$DCC0, T1, T2$
B7	$\overline{DCC1}$	B15	$DCC1, T0, T3$

4.3.1.5 Row Addressing

Although the B-group has only eight physical rows, it contains 16 reserved addresses, B0–B15. Table 4.1 lists the 16 addresses and the corresponding word-lines. Among 16 addresses, the first four addresses, B0–B3, are designated for operands (i.e., T0–T3), and the next four addresses, B4–B7, are assigned to two DCC rows for bulk NOT operation. B8–B11 activates two wordlines simultaneously for the copy. For example, B8 activates T0 and n-wordline of DCC0 to move negated values to DCC0. B12–B15 actives three wordlines together for bulk bitwise AND/OR operation. For example, B12 activates T0, T1, and T2 at the same time, and the computed value will be re-written to all the rows.

4.3.1.6 AMBIT Command Execution

To execute the proposed bulk bitwise operation, which involves row-wide data copies and logical computation, AMBIT supports a fused complex command primitive named AAP (Activate–Activate–Pre-charge). By combining row activate, row activate, and pre-charge back to back, the AAP primitive reduces the number of required commands significantly. Hence, it reduces the total latency. Figure 4.7 shows how the basic logical operations can be done with the AAP primitive. AAP (Di, B0) means that it activates Di from the D-group and activates B0 from the B-group to copy Di to T0 and pre-charges for the following command. Likewise, AAP (Dj, B1) copies the Dj from the D-group to T1. AAP(C0, B2) sets the T2 to 0 for bulk AND operation. Finally, AAP(B12, Dk) executes TRA with B12, and the final AND result will be copied to Dk in the D-group.

4.3.1.7 Evaluation

To evaluate the AMBIT architecture, the authors compare the raw throughput of bulk bitwise operations of AMBIT, such as NOT, AND/OR, NAND/NOR, and XOR/XNOR, against Intel Skylake CPU and NVIDIA GeForce GTX 745 GPU. As expected, AMBIT outperforms the others with DDR3 memories by 32x

Dk = Di **and** Dj	Dk = Di **xor** Dj = (Di & !Dj) | (!Di & Dj)
AAP (Di, B0) ;T0 = Di AAP (Dj, B1) ;T1 = Dj AAP (C0, B2) ;T2 = 0 AAP (B12, Dk) ;Dk = T0 & T1	
Dk = Di nand Dj	AAP (Di, B8) ;DCC0 = !Di, T0 = Di AAP (Dj, B9) ;DCC1 = !Dj, T1 = Dj AAP (C0, B10) ;T2 = T3 = 0
AAP (Di, B0) ;T0 = Di AAP (Dj, B1) ;T1 = Dj AAP (C0, B2) ;T2 = 0 AAP (B12, B5) ;DCC0 = !(T0 & T1) AAP (B4, Dk) ;Dk = DCC0	AAP (B14) ;T1 = DCC0 & T1 AAP (B15) ;T0 = DCC1 & T0 AAP (C1, B2) ;T2 = 1 AAP (B12, Dk) ;Dk = T0 | T1

Fig. 4.7 Programming with AAP command primitive

improvement in throughput and 44x reduction in energy consumption. For a real-world application such as database bitmap indices, which utilizes bulk bitwise operations a lot, AMBIT accelerates the baseline CPU operation by 6x on average.

4.3.2 DRISA

4.3.2.1 Motivation

The goal of DRISA is to merge the strength of memory-rich processors such GPUs and ASIC-based neural processing units (NPUs) [9] and compute-capable PIMs [1]. The memory-rich processors show high performances using abundant memory bandwidth but have little memory capacity. On the other hand, compute-capable PIMs suffer from low performance. To have both strengths, DRISA builds a PIM accelerator based on DRAM technology. Like AMBIT, DRISA adds the logic operations at the level of bitline sense amplifiers to leverage the maximal internal bandwidth while minimizing the design changes from regular DRAM.

Figure 4.8 shows the overview of DRISA. The highlighted regions with green and blue depict the building blocks that require design changes from a regular DRAM. At the chip level, DRISA modifies the group and bank buffers to facilitate internal data transfers. It also modifies the bank controller to control logic processing in multiple subarrays in each bank. At the lowest cell matrix level, it adds logic gates and shifters at the bottom of the DRAM cells.

4.3.2.2 Cell Microarchitectures

Unlike AMBIT uses the same DRAM cell architecture as the regular DRAMs for feasibility/manufacturability, DRISA proposes three different DRAM cell architectures: 3T1C, 1T1C-NOR/MIX, and 1T1C-ADDER (Fig. 4.9).

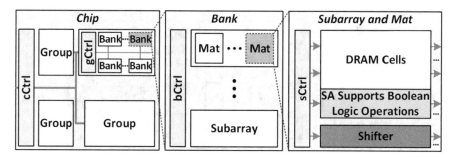

Fig. 4.8 DRISA overview and design changes

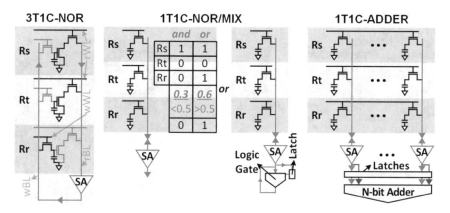

Fig. 4.9 Three DRAM cells of DRISA

1T1C-NOR/MIX adds a NOR gate or other gates below each bitline sense amplifier with a latch. It performs bitwise logic operation between the read operand and the latched operand. On the other hand, 1T1C-ADDER adds the latches and a parallel adder below multiple sense amplifiers. However, both of them are difficult to be realized considering the extremely narrow DRAM cell pitch. The simplest case requires 4 transistors for a NOR gate and 8 transistors for a latch. Having to route metal connections as well, it is not trivial to integrate the logic within the DRAM cell pitch, even for the simplest case.

The 3T1C cell, illustrated in more detail in Fig. 4.10, was used in early DRAM design. It has separated wordlines, one for write and the other for read operation, and two transistors to connect them (M1 and M3). The M2 transistor decouples the other two transistors, and its gate is connected to the cell capacitor. If the M3 transistor is enabled, it is connected to the read bitline BL2 with having the cell capacitor as the input value. From the bitline perspective, M2 transistors connected in parallel implement NOR operation. In other words, if only a single cell value is 1, the bitline value will go to the ground. Using this native NOR configuration of the 3T1C cell, DRISA can perform bulk bitwise NOR operations between the two

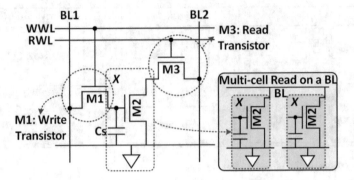

Fig. 4.10 3T1C cell for native NOR operation

Fig. 4.11 NOR-based
selector logic implementation

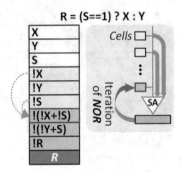

rows. When it activates two wordlines through M3 transistors simultaneously, the
bitlines will eventually have the NOR result between the two rows.

4.3.2.3 Computing Using NOR Operation

While AMBIT utilizes bitwise AND/OR and NOT operation to implement logic
functions, DRISA only uses bitwise NOR operation as it is functionally complete.
As explained in the previous section, DRISA activates two rows simultaneously
to compute bitwise NOR operation using the native NOR connection of the 3T1C
cells. Figure 4.11 illustrates the DRISA's NOR-based selector, or multiplexer, logic
implementation. The Boolean equation for the selector is $R = SX + \sim SY$,and this
can be re-written as $R = \sim NOR(NOR(\sim S, \sim X), NOR(S, \sim Y))$ using 3 NOR
operations and 4 NOT operations. NOT operation can also be computed using NOR
by having one of the input operands to 0 $(NOT(X) = NOR(0, X))$. As a result,
the bulk bitwise selector logic can be done in 7 steps in DRISA.

As the bulk bitwise operation applies the same low-level logic operations to all
the bits, mapping higher-level logic functions is not straightforward. To address
this problem a little bit, DRISA includes shifters under the bitline sense amplifiers
for data communication among neighbor bitlines. As a simple but essential use

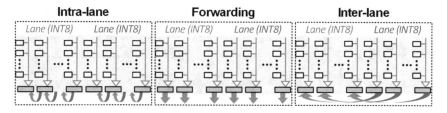

Fig. 4.12 Shifting operations in DRISA

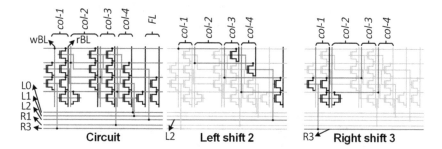

Fig. 4.13 Transistor-level shifter circuits

case, the shifter can propagate carry-out signals to the neighbor bitlines in addition. Specifically, DRISA supports three types of shifting operations: intra-lane, inter-lane, and forwarding, as shown in Fig. 4.12. As the name implies, intra-lane shift is a single-bit shift to the neighbor bitlines inside the lane, and inter-lane shift is a shift in a lane unit such as byte shift or word shift. The lane means a unit of data, such as 8 bits or 16 bits. Forwarding is just a read without any shift applied.

As the shifter implementation can be complex, DRISA proposes transistor-level shifter circuits. Figure 4.13 shows the 4-bit intra-lane shifter circuits with the example of left shift by 2 and right shift by 3, where rBL, wBL, and FL means read bitline, write bitline, and filling line, respectively. According to the control lines in the bottom (L0, L1, L2, R1, and R3), only the necessary transistors are enabled for barrel shifting operations. For example, the read bitlines of columns 3 and 4 are enabled, and the read values are transferred to columns 1 and 2, respectively, for the left shift by 2.

4.3.2.4 Evaluation

To evaluate the proposed processing-in-memory architecture, the authors heavily modify CACTI-3DD [10] for circuit-level simulation. CACTI-3DD is a highly accurate circuit-level DRAM simulator that provides DRAM's latency, energy consumption, and area. They also use Synopsys's Design Compiler for the controllers and adders in the 1T1C-adder configuration. They use the logic process simply

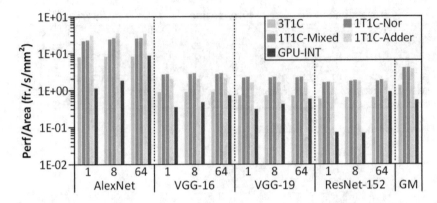

Fig. 4.14 DRISA experimental results

because the DRAM process's technology information is unavailable. Instead, they scale the two different technologies based on an old technical paper [11], which states the DRAM process is 22% slower and 80% bigger than a comparable logic process. However, we cannot be convinced that this scaling is still valid as the two processes have been differed a lot since then. They also create an in-house behavior-level performance simulator from scratch, which evaluates DRISA's latency and power consumption for a given task. In addition to memory organization changes, the performance simulator gets neural network topology and mapping configurations as input.

Figure 4.14 shows the area-normalized performance results of the various DRISA configurations against the Nvidia Titan X GPU. It measures the number of frames per second on CNN models such as AlexNet [12], VGG-16 [13], VGG-19, and ResNet-152 [14] with batch sizes of 1, 8, and 64. Interestingly, 3T1C case is not good because of its large memory cell area. 1T1C-ADDER is also not the best because the computing and data movement are not balanced. Out of this experiment, 1T1C-MIXED case that attaches NAND, NOR, XNOR, and INV gate to a sense amplifier performs best, thanks to its logic coverage.

4.4 Bank-Level Processing-in-Memory

The bulk bitwise operation that implements logic at the bitline sense amplifier level is the best in terms of internal data bandwidth. However, area constraint for this method is the toughest because of the narrow cell pitch; the cell pitch has kept decreasing to integrate more cells, having a high capacity. The next possible level is bank level, which integrates processing logic after column decoder and selector. Since the processing logic can enjoy the whole width of the cell array, not a single cell pitch, it is more affordable to add logic functions in the space. In addition, as every commercial DRAM has column selectors, this method is less invasive as it does not change any design at the cell array matrix. Also, it utilizes the possible maximum bandwidth of the existing DRAM architectures.

4.4.1 Newton

4.4.1.1 Motivation

Newton is a feasible accelerator-in-memory architecture in a commercial DRAM proposed by one of the major DRAM makers, SK Hynix. Since they know that the DRAM process requires more-than-expected area in making logic gates, they chose to integrate logic at the bank level, i.e., after column select, rather than at the bitline sense amplifier level. By doing this, Newton gains enough physical space for logic integration, while it loses internal bandwidth.

Rather than targeting generic DNN workloads, Newton tries to find a good match in applications, which its PIM architecture can address well. It focuses on memory-bound deep learning models such as language models (e.g., Google's BERT [15] and OpenAI's GPT [16]) and recommendation systems (e.g., Facebook's DLRM [17]). These models are easily bottlenecked by memory read bandwidth because they have huge model sizes with low-data reuse opportunities, mainly caused by massive matrix-vector multiplications. Therefore, the PIM architecture that supports sufficient internal read operations with lightweight logic processing can be very effective for this type of models. The opposite case will be deep CNN models that require more computations per data read with a high-data reuse opportunity. In conclusion, Newton targets the deep learning models with fully connected layers for a single-batched inference scenario.

4.4.1.2 Architecture

Figure 4.15 shows the overall architecture of Newton in a single DRAM die. It has a total of 16 banks where each bank includes 16 multipliers, 16 adders, and 16-bit accumulation register. As mentioned earlier, it integrates the above logic gates after the column decoder, i.e., 32:1 column mux in the diagram, to make it feasible with minimizing the changes in the memory bank design. Unlike AMBIT or DRISA, it does not have to change the design in row decoder and row drivers because it does not require multi-row activation.

Newton activates a single row in a bank, like in a normal DRAM, which has the size of 1KB. Among them, only 32 bytes are selected after 32:1 column select. As Newton uses the half-precision floating-point data type (FP16), it reads 16 FP16 data at a time out of 512 FP16 data in a row. Each of FP16 data enters to an input of each multiplier. The other input comes from the global buffer. The multipliers multiply between the 16 FP data from the global buffer and the 16 FP data from the

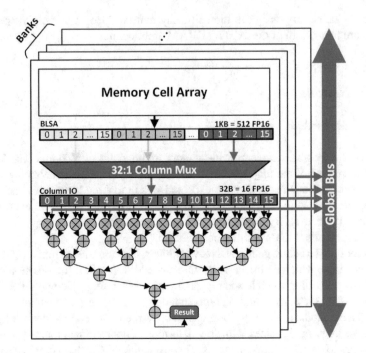

Fig. 4.15 Newton architecture

bank, and the 16 products are accumulated through the adder tree to a single FP16 result.

In Newton, the global buffer broadcasts an input vector to the memory banks, while the banks store different parts of the weight matrix, as illustrated in Fig. 4.16. The large weight matrix is chunked into tiles, whose size is 16 rows by 512 FP16 data, and the rows in a tile are interleaved over the multiple banks. The input vector is also segmented into the groups of 512 FP16 data, and they are distributed to the banks for matrix-vector operation. To increase the internal read bandwidth, Newton activates multiple banks at the same time. This multi-bank activation, or bank-level parallelism, is a key differentiator from a regular DRAM. It increases the internal read bandwidth and, hence, its compute bandwidth.

4.4.1.3 Newton's Operation

Figure 4.17 shows the overall operation of Newton, including new commands and multi-bank activation for PIM operations. First, it loads the global buffer with the input vector data using GWRITE command. Then, it activates multiple banks using G_ACT command. Although activating all 16 banks would be the best option for achieving high throughput, it is difficult because of power and voltage drop issue. In this chapter, Newton can activate four banks at a time and needs an interval time

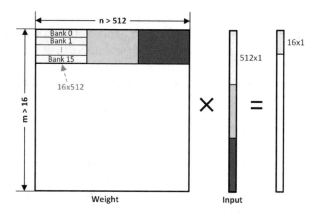

Fig. 4.16 Data mapping in Newton

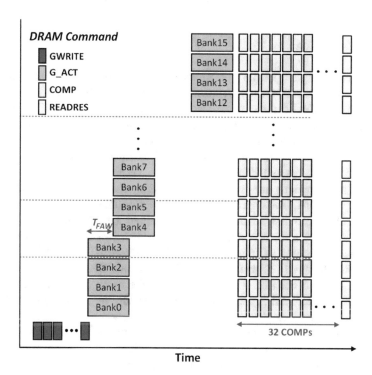

Fig. 4.17 Newton's PIM operations

to the next multi-bank activation. This interval time is defined as four-activation timing window or T_{FAW}. After issuing G_ACT command four times to activate all the 16 banks, Newton performs in-bank multiply-and-accumulate (MAC) operations using COMP command. The COMP command applies to all the banks and repeats 32 times to cover an entire row, noting that Newton integrates logic gates after the 32:1 column select. To cover a complete row during the global bank activation, it needs to repeat the COMP command. After the COMP commands, each bank should have the final accumulation result. Using READRES command, the top controller gathers all the results from the banks.

To reduce the time for an overall PIM operation, reducing the interval between multi-bank activations, or T_{FAW}, is important. As DRAM already has several internal voltage domains with lots of transistor loads under them, it is quite challenging to activate multiple banks simultaneously because it causes a severe internal voltage drop. To have a small T_{FAW}, internal low-dropout (LDO) regulator and DC–DC pump driver should be designed to provide enough strength.

4.4.1.4 Evaluation

Newton evaluates its speedup performance by comparing ideal non-PIM, non-optimized Newton, and Newton architecture to Nvidia Titan V GPU. Ideal non-PIM architecture belongs to the conventional von Neumann architecture but with the assumption of unlimited computed bandwidth. Its only bottleneck is the DRAM's off-chip memory bandwidth. While, non-optimized Newton architecture is used as a reference model with five Newton features eliminated, described in previous sections: all-bank ganged command (*ganged*), complex multi-step commands (*complex*), reuse input by interleaved layout and tiling mechanism (*reuse*), four-bank activation (*four bank*), and improved T_{FAW}. Newton architecture and non-optimized Newton architecture are simulated using modified DRAMSim2 [18], and the ideal non-PIM is simulated using GPGPU-Sim [19]. Newton uses matrix-vector multiplication benchmarks from GNMT LSTM [20], BERT [15], fully connected layers from AlexNet [12], and DLRM [17]. Newton's DRAM configuration for simulation is set to HBM2E. Figure 4.18 shows the result of speedup over GPU. Newton proves that PIM is indispensable by showing the limitations of GPU in computing memory-bound applications. The result shows that Newton achieves 54× speedup over Titan-V GPU on average across individual layers, while ideal non-PIM achieves only up to 5.4× speedup. Additionally, Newton discusses the importance of the optimization in PIM commands. Figure 4.19 shows the analysis result of non-optimized Newton architectures. Starting from non-opt Newton without any features, each feature is gradually added in a given order appeared in the graph one by one. Non-optimized Newton shows 48% speedup over the Titan-V GPU, while the Newton with all the optimization schemes offers 54× speedup. Especially, the *gang* commands increase the performance the most with its all-bank operation that reduces the command bandwidth by 16×. The *complex* commands additionally reduce command bandwidth by 3×.

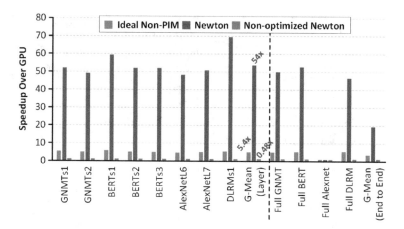

Fig. 4.18 Newton's speedup evaluation

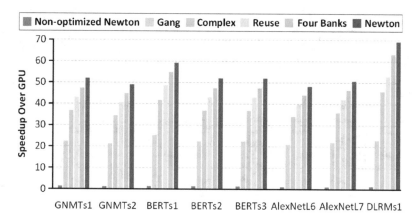

Fig. 4.19 Isolating Newton's optimization

4.4.2 HBM-PIM

4.4.2.1 Motivation

HBM-PIM is the first fabricated stacked-DRAM-based PIM solution from Samsung, one of the major memory vendors. As shown in Fig. 4.20, it has 8 DRAM channels with a total of 1024 pins and integrates programmable compute units in 4 out of 8 DRAM dies. Each DRAM die includes 16 banks with 128 IO pins, where each bank's capacity is 64MB (32MB with PIM) and each pin's bandwidth is 2.4 Gbps. Unlike Newton, which includes the logic with the fixed function, HBM-PIM includes the logic with a bit of programmability to address the computational requirements of AI applications.

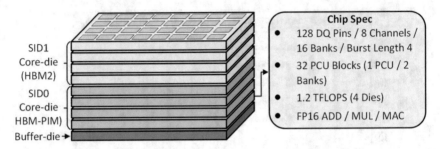

Chip Spec

- 128 DQ Pins / 8 Channels / 16 Banks / Burst Length 4
- 32 PCU Blocks (1 PCU / 2 Banks)
- 1.2 TFLOPS (4 Dies)
- FP16 ADD / MUL / MAC

Fig. 4.20 HBM-PIM overview

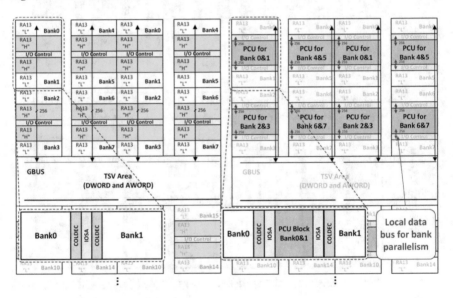

Fig. 4.21 HBM-PIM architecture

4.4.2.2 HBM-PIM Architecture

Figure 4.21 shows the overall architecture of HBM-PIM, illustrating how it is evolved from the existing HBM2 design. The left diagram shows the architecture of a single DRAM die used in the conventional HBM2. It has 16 banks in total, where each bank has a row decoder and a column decoder, while the two banks in the top and bottom share the IO sense amplifiers. Four banks make up a bank group, and the bank group shares the 256-bit bank group bus, which eventually muxed into the global bus. Since this is 3-d stacked memory, the global bus goes down to the base die through the through-silicon-via (TSV) area.

On the right side, the HBM-PIM integrates a programmable compute unit (PCU) per two banks. Each bank pair now shares the PCU with the separate IO sense amplifiers. It also introduces the 256-bit local data buses that interconnect each

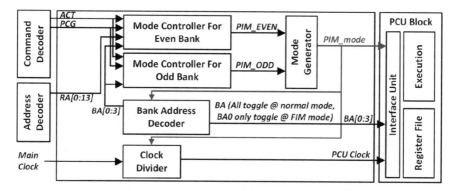

Fig. 4.22 Internal PIM controller

bank and the shared PCU. In addition, the PIM controller for operating the PCUs is integrated in the TSV area. To maximize its internal bandwidth, HBM-PIM uses the bank-level parallelism, as Newton does. As two banks share one compute unit to limit the power and area cost, it activates half of the banks (i.e., even banks or odd banks) in the die at the same time.

4.4.2.3 HBM-PIM Controller

From the host side, the proposed HBM-PIM is seen exactly the same as a regular HBM, being compatible with the existing DRAM interfaces. With the PIM instructions stored in the command register file in the PCU, the host can control every PIM instruction with conventional load and store instructions to specific memory addresses. The only thing the controller inside the DRAM should do is the mode change between the normal and PIM.

As shown in Fig. 4.22, the internal PIM controller decodes specific combinations between the command and address to generate a mode change signal. For example, if the row activate (ACT) command comes with a specific address in bank 0, PIM_Even signal is asserted. Likewise, if the same command and an address come in for bank 1, PIM_Odd signal is asserted. Only if both signals are asserted, the mode changes to the PIM mode. With the start of PIM mode, the PCU gets its clock to run the PIM instructions. Once the PIM operations finish, the controller changes the PIM mode to normal, again with the combination of a command and a specific address.

4.4.2.4 Programmable Computing Unit

In the execution model, the major difference between Newton and HBM-PIM is that Newton adds a few special PIM commands to use the integrated arithmetic units

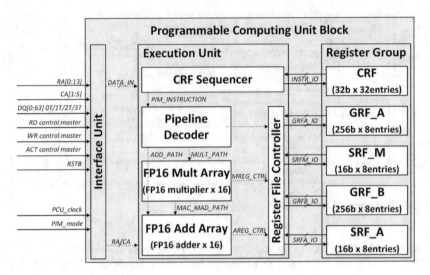

Fig. 4.23 Programmable computing unit

with a fixed function, while HBM-PIM changes the mode to activate the compute unit named PCU. PCU is a programmable unit with its own instructions. Figure 4.23 shows the block diagram of the PCU, consisting of an interface unit, execution unit, and register group. The interface unit receives control and data signals from the memory's command controller. The execution unit includes a pair of 16 FP16 multipliers and adders. Each of them has a 5-stage pipeline and works in parallel with single-instruction multiple-data (SIMD) fashion.

The register group includes the command register file (CRF), general-purpose register file (GRF), and scalar register file (SRF). The CRF buffers up to 32 32-bit PIM instructions. The GRF composed of sixteen 256-bit registers is evenly divided into GRF_A and GRF_B for even bank and odd bank, respectively. The SRF replicates a scalar value to a vector and performs scalar multiplications or scalar additions to a source operand from GRF. Like other in-order cores, the PCU fetches a PIM instruction from the CRF, decodes it, reads source operands to the SIMD FP units, and stores the result back to the GRF. Table 4.2 shows the overall 9 PIM instructions.

4.4.2.5 Operation Flow

HBM-PIM has three operational steps. First, the host stores input data to the DRAM cell arrays. As it is set to normal mode with initialization, the host accesses each bank as a regular DRAM. Then, the host changes the operation mode from normal to PIM. Second, the host sends instructions and weight data to PCUs via the DQ interface. The PCU can save up to 32 instructions in the CRF, and its program counter reads the instructions one by one from address 0. Third, once each PCU

Table 4.2 PCU instructions

Type	Command	Description
Floating point	ADD	FP 16 addition
	MUL	FP 16 multiplication
	MAC	FP 16 multiply-and-accumulate
	MAD	FP 16 multiply and add
Data path	MOVE	Load or store data
	FILL	Copy data from bank to GRFs
Control path	NOP	Do nothing
	JUMP	Jump instruction
	EXIT	Exit instruction

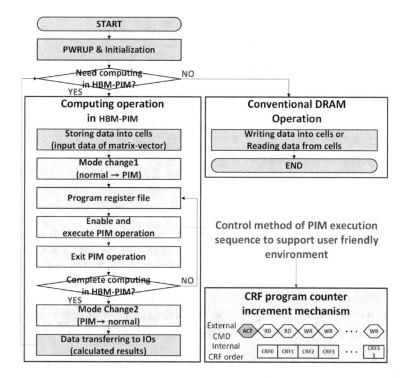

Fig. 4.24 HBM-PIM operations

completes the PIM operations such as matrix-vector multiplication, it transfers the results in the GRF to the DRAM cell arrays. Note that it is done across multiple banks (even or odd banks). The host finally switches the mode back to normal and reads the results from each bank. The flow chart in Fig. 4.24 summarizes the overall operation flow of HBM-PIM.

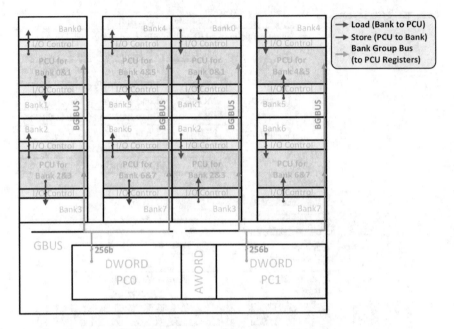

Fig. 4.25 Data buses in HBM-PIM

4.4.2.6 Data Movements

As HBM-PIM introduces PCUs between even and odd banks, it needs data buses to transfer data among them. The local data buses are responsible for data transfers between PCUs and banks. With the MOVE command, the PCU can load data from the cell array to the GRF or store data from the GRF to the cell array. This is a multi-bank operation; either even or odd banks can be enabled simultaneously. The host uses the bank group global bus to issue instructions and weight data to the PCUs. Figure 4.25 depicts the two types of buses for data movements in HBM-PIM.

4.4.2.7 Implementation Results

HBM-PIM is the first PIM chip ever fabricated in HBM using a 20 nm DRAM process. Figure 4.26 shows the chip micrograph and measurement results. It achieves 2.4 Gbps/pin operation, without power consumption increase from HBM2, and PCU operation at 300 MHz. In addition, an FPGA-based test platform and an emulation environment confirm the system performance can improve by $2.1\times$ for DeepSpeech2 benchmark [21] while reducing the system energy by 71% compared to a typical GPU system using HBM2.

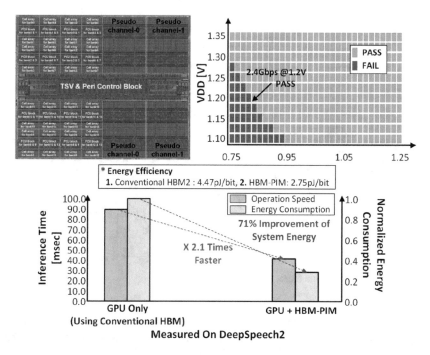

Fig. 4.26 HBM-PIM chip photo and measurement results

4.5 3-D Processing-in-Memory

In this section, we look into the PIM architectures using full 3-d vertical stacking of memory and logic. The HBM-PIM that we describe in Sect. 4.4 is a 2.5-d solution; it puts the logic module and HBM side by side via silicon interposer. On the other hand, the full 3-d stacking means the integration of the logic die in the bottom with the stacked memories like an HBM on top of it. Since it stacks the main compute die and stacked memory dies, it is further advanced from HBM, expecting more energy-efficient data communications between the two entities. Hybrid memory cube (HMC) is the main example of this. However, the realization of 3-d PIM can be difficult due to tight physical and timing constraints among 3-d stacked dies. All the proposed 3-d PIM architectures are evaluated only using simulation. In this section, we briefly review a few works based on the 3-d PIM architecture.

4.5.1 Neurocube

Neurocube [5] is one of the earliest architectures that demonstrates the feasibility and performance benefits of using a 3-d high-density memory package for deep

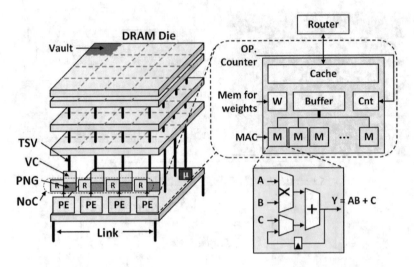

Fig. 4.27 Neurocube architecture

neural networks. As shown in Fig. 4.27, the Neurocube architecture is designed in the logic die of an HMC, consisting of the global controller, programmable neurosequence generators (PNGs), and processing elements (PEs) connected by a network-on-chip. While, each PE is assigned for a set of vertically connected DRAM banks called vault. The PE is the main computing unit having multiple multiply-and-accumulate (MAC) units to accelerate deep neural network computations. The PNG generates a correct sequence of data accesses to the vault using the vault controller and pushes them into the MAC units. The network-on-chip with 2-d mesh topology interconnects all the PEs to enable inter-vault communications for various data mappings and operations.

4.5.2 Tetris

Tetris [6] architecture is based on Neurocube architecture. The HMC stack is vertically divided into sixteen 32-bit-wide vaults, in which each vault functions as a channel that controls all the banks inside (two banks per die). Different from Neurocube, Tetris chooses an array design for processing units. Each processing unit includes a small global on-chip buffer to maximize data reuse opportunities. Tetris also proposes a scheduling algorithm for efficient data flow. Focusing on input data reuse, it buffers the input feature maps on the global buffer with tiling and streams the output feature maps and weight filters directly to/from the external memory. Figure 4.28 illustrates Tetris architecture.

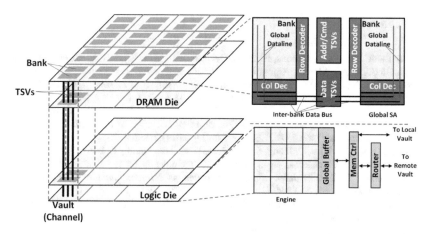

Fig. 4.28 Tetris architecture

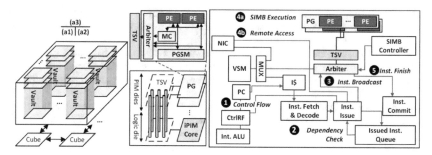

Fig. 4.29 iPIM architecture

4.5.3 iPIM

iPIM [7] compromises the 3-d PIM approach that Neurocube and Tetris used and the bank-level PIM approach that Newton and HBM-PIM used, in order to increase effective compute bandwidth and reduce energy spent from data movements via TSVs. As Fig. 4.29 shows, iPIM's vault architecture decouples the role to control and execution. The logic die includes the iPIM core that performs complex control operations such as instruction decoding and issuing, and memory bank controls. On the other hand, the process group (PG), integrated into each DRAM die of a vault, performs simple but memory-intensive operations at near bank. To enable massive bank-level concurrent execution, iPIM proposes single-instruction multiple-bank (SIMB) instructions, including computation, index calculation, intra/inter-vault data movement, and synchronization operation.

References

1. V. Seshadri, D. Lee, T. Mullins, H. Hassan, A. Boroumand, J. Kim, M.A. Kozuch, O. Mutlu, P.B. Gibbons, T.C. Mowry, Ambit: In-memory accelerator for bulk bitwise operations using commodity DRAM technology, in *2017 50th Annual IEEE/ACM International Symposium on Microarchitecture (MICRO)*. IEEE, Piscataway (2017), pp. 273–287
2. S. Li, D. Niu, K.T. Malladi, H. Zheng, B. Brennan, Y. Xie, DRISA: a DRAM-based reconfigurable in-situ accelerator, in *2017 50th Annual IEEE/ACM International Symposium on Microarchitecture (MICRO)*. IEEE, Piscataway (2017), pp. 288–301
3. M. He, C. Song, I. Kim, C. Jeong, S. Kim, I. Park, M. Thottethodi, T.N. Vijaykumar, Newton: a DRAM-maker's accelerator-in-memory (AiM) architecture for machine learning, in *2020 53rd Annual IEEE/ACM International Symposium on Microarchitecture (MICRO)*. IEEE, Piscataway (2020), pp. 372–385
4. S. Lee, S.-h. Kang, J. Lee, H. Kim, E. Lee, S. Seo, H. Yoon, S. Lee, K. Lim, H. Shin, J. Kim, O. Seongil, A. Iyer, D. Wang, K. Sohn, N.S. Kim, Hardware architecture and software stack for PIM based on commercial DRAM technology: Industrial product, in *2021 ACM/IEEE 48th Annual International Symposium on Computer Architecture (ISCA)*. IEEE, Piscataway (2021), pp. 43–56
5. D. Kim, J. Kung, S. Chai, S. Yalamanchili, S. Mukhopadhyay, Neurocube: a programmable digital neuromorphic architecture with high-density 3D memory. ACM SIGARCH Comput. Archit. News **44**(3), 380–392 (2016)
6. M. Gao, J. Pu, X. Yang, M. Horowitz, C. Kozyrakis, TETRIS: scalable and efficient neural network acceleration with 3d memory, in *Proceedings of the Twenty-Second International Conference on Architectural Support for Programming Languages and Operating Systems* (2017), pp. 751–764
7. P. Gu, X. Xie, Y. Ding, G. Chen, W. Zhang, D. Niu, Y. Xie, iPIM: programmable in-memory image processing accelerator using near-bank architecture, in *2020 ACM/IEEE 47th Annual International Symposium on Computer Architecture (ISCA)*. IEEE, Piscataway (2020), pp. 804–817
8. V. Seshadri, Y. Kim, C. Fallin, D. Lee, R. Ausavarungnirun, G. Pekhimenko, Y. Luo, O. Mutlu, P.B. Gibbons, M.A. Kozuch, T.C. Mowry, RowClone: Fast and energy-efficient in-DRAM bulk data copy and initialization, in *Proceedings of the 46th Annual IEEE/ACM International Symposium on Microarchitecture*, (2013), pp. 185–197
9. N.P. Jouppi, C. Young, N. Patil, D. Patterson, G. Agrawal, R. Bajwa, S. Bates, S. Bhatia, N. Boden, R.B. Al Borchers, P.-l. Cantin, C. Chao, C. Clark, J. Coriell, M. Daley, M. Dau, J. Dean, B. Gelb, T.V. Ghaemmaghami, R. Gottipati, W. Gulland, R. Hagmann, C.R. Ho, D. Hogberg, J. Hu, R. Hundt, D. Hurt, J. Ibarz, A. Jaffey, A. Jaworski, A. Kaplan, H. Khaitan, D. Killebrew, A. Koch, N. Kumar, S. Lacy, J. Laudon, J. Law, D. Le, C. Leary, Z. Liu, K. Lucke, A. Lundin, G. MacKean, A. Maggiore, M. Mahony, K. Miller, R. Nagarajan, R. Narayanaswami, R. Ni, K. Nix, T. Norrie, M. Omernick, N. Penukonda, A. Phelps, J. Ross, M. Ross, A. Salek, E. Samadiani, C. Severn, G. Sizikov, M. Snelham, J. Souter, D. Steinberg, A. Swing, M. Tan, G. Thorson, B. Tian, H. Toma, E. Tuttle, V. Vasudevan, R. Walter, W. Wang, E. Wilcox, D.H. Yoon, In-datacenter performance analysis of a tensor processing unit, in *Proceedings of the 44th Annual International Symposium on Computer Architecture*, (2017), pp. 1–12
10. K. Chen, S. Li, N. Muralimanohar, J.H. Ahn, J.B. Brockman, N.P. Jouppi, CACTI-3DD: architecture-level modeling for 3D die-stacked DRAM main memory, in *2012 Design, Automation & Test in Europe Conference & Exhibition (DATE)*. IEEE, Piscataway (2012), pp. 33–38
11. Y.B. Kim, T.W. Chen, Assessing merged DRAM/logic technology. Integration **27**(2), 179–194 (1999)
12. A. Krizhevsky, I. Sutskever, G.E. Hinton, ImageNet classification with deep convolutional neural networks. Adv. Neural Inform. Process. Syst. **25**, 1097–1105 (2012)
13. K. Simonyan, A. Zisserman, Very deep convolutional networks for large-scale image recognition (2014). arXiv preprint arXiv:1409.1556

14. K. He, X. Zhang, S. Ren, J. Sun, Deep residual learning for image recognition, in *Proceedings of the IEEE Conference on Computer Vision and Pattern Recognition* (2016), pp. 770–778
15. J. Devlin, M.W. Chang, K. Lee, K. Toutanova, BERT: pre-training of deep bidirectional transformers for language understanding (2018). arXiv preprint arXiv:1810.04805
16. OpenAI, GPT-3 powers the next generation of apps, In: *OpenAI* (2021). https://openai.com/blog/gpt-3-apps/. Accessed 5 Nov 2021
17. M. Naumov, D. Mudigere, H.-J.M. Shi, J. Huang, N. Sundaraman, J. Park, X. Wang, U. Gupta, C.-J. Wu, A.G. Azzolini, D. Dzhulgakov, A. Mallevich, I. Cherniavskii, Y. Lu, R. Krishnamoorthi, A. Yu, V. Kondratenko, S. Pereira, X. Chen, W. Chen, V. Rao, B. Jia, L. Xiong, and M. Smelyanskiy, Deep learning recommendation model for personalization and recommendation systems (2019). arXiv preprint arXiv:1906.00091
18. P. Rosenfeld, E. Cooper-Balis, B. Jacob, DRAMSim2: a cycle accurate memory system simulator. IEEE Comput. Archit. Lett **10**(1), 16–19 (2011)
19. A. Bakhoda, G.L. Yuan, W.W. Fung, H. Wong, T.M. Aamodt, Analyzing CUDA workloads using a detailed GPU simulator, in *2009 IEEE International Symposium on Performance Analysis of Systems and Software*. IEEE, Piscataway (2009), pp. 163–174
20. Y. Wu, M. Schuster, Z. Chen, Q.V. Le, M. Norouzi, W. Macherey, M. Krikun, Y. Cao, Q. Gao, K. Macherey, J. Klingner, A. Shah, M. Johnson, X. Liu, Ł. Kaiser, S. Gouws, Y. Kato, T. Kudo, H. Kazawa, K. Stevens, G. Kurian, N. Patil, W. Wang, C. Young, J. Smith, J. Riesa, A. Rudnick, O. Vinyals, G. Corrado, M. Hughes, J. Dean, J. Google's neural machine translation system: bridging the gap between human and machine translation (2016). arXiv preprint arXiv:1609.08144
21. D. Amodei, S. Ananthanarayanan, R. Anubhai, J. Bai, E. Battenberg, C. Case, J. Casper, B. Catanzaro, Q. Cheng, G. Chen, J. Chen, J. Chen, Z. Chen, M. Chrzanowski, A. Coates, G. Diamos, K. Ding, N. Du, E. Elsen, J. Engel, W. Fang, L. Fan, C. Fougner, L. Gao, C. Gong, A. Hannun, T. Han, L. Johannes, B. Jiang, C. Ju, B. Jun, P. LeGresley, L. Lin, J. Liu, Y. Liu, W. Li, X. Li, D. Ma, S. Narang, A. Ng, S. Ozair, Y. Peng, R. Prenger, S. Qian, Z. Quan, J. Raiman, V. Rao, S. Satheesh, D. Seetapun, S. Sengupta, K. Srinet, A. Sriram, H. Tang, L. Tang, C. Wang, J. Wang, K. Wang, Y. Wang, Z. Wang, Z. Wang, S. Wu, L. Wei, B. Xiao, W. Xie, Y. Xie, D. Yogatama, B. Yuan, J. Zhan, Z. Zhu, Deep speech 2: End-to-end speech recognition in English and mandarin, in *International Conference on Machine Learning* (2016), pp. 173–182. PMLR

Chapter 5
ReRAM-Based Processing-in-Memory (PIM)

Tony Tae-Hyoung Kim, Lu Lu, and Yuzong Chen

5.1 Introduction

Emerging applications such as artificial intelligence and machine learning have created interest in hardware accelerators for processing parallel data considerably in various neural networks. One of the most critical arithmetic functions in neural networks is Multiply-and-Accumulate (MAC). In the conventional computing architecture (i.e., Von Neumann architecture), processing elements and memory are separated. MAC operations require massive data transfer between processing elements and memory, which consumes a huge amount of power. Recently, processing-in-memory (PIM) architectures have been introduced to address the bottleneck of Von Neumann architecture [1–10]. Since PIM architectures include local computing circuits and memory, we can minimize the data transfer from/to external memory. In general, it is well known that the PIM architectures can improve energy efficiency by orders of magnitude. While SRAM and DRAM are commonly considered in PIM architectures, ReRAM has also gained increasing interest because of the accelerator development for edge computing [11–17]. Many edge devices such as wearables, IoT devices, and biomedical devices require high energy efficiency with compromised performance. Particularly, the edge devices process data scarcely and mostly stay in the standby condition. Since ReRAM offers moderate performance and non-volatility, ReRAM-based PIM architectures can be promising solutions for edge computing. However, ReRAM technologies are not mature yet, and various design issues for ReRAM-based PIM should be tackled comprehensively. In this chapter, we will introduce various ReRAM-based

T. T.-H. Kim (✉) · L. Lu · Y. Chen
Nanyang Technological University, Singapore, Singapore
e-mail: thkim@ntu.edu.sg; Lu_Lu@ime.a-star.edu.sg; yuzong@nus.edu.sg

© The Author(s), under exclusive license to Springer Nature Switzerland AG 2023
J.-Y. Kim et al. (eds.), *Processing-in-Memory for AI*,
https://doi.org/10.1007/978-3-030-98781-7_5

PIM designs covering from unit cells, circuit techniques, and architectures. Besides MAC, other essential logic-in-memory functions will also be discussed.

5.2 Basic ReRAM PIM Operation

Figure 5.1 illustrates a typical ReRAM array for PIM. In general, the ReRAM array for PIM looks the same as the ReRAM array for regular memory operation. Since the 1T1R ReRAM cell is compact and can provide high density, many ReRAM PIM accelerators are designed using the typical 1T1R ReRAM cell. The key difference between normal ReRAM and ReRAM PIM is the number of rows that are activated simultaneously. In normal ReRAM, only one row is accessed at a given time for programming and read operation. However, ReRAM PIM accesses multiple rows for reading operation, which is the most frequently executed operation for neural networks. Activating multiple rows will create multiple current components in each bitline, which is used as an analog multiply-and-accumulate (MAC) result. The analog MAC output can be represented as follows:

$$MAC = \sum_{i=0}^{v} IN_i \times W_i = \sum_{i=0}^{v} I_{MC[i]}$$

Here, IN_i, W_i, $I_{MC[i]}$, and v are input signal, weight, cell current, and the number of selected rows, respectively. IN_i affects WLs in Fig. 5.1 and is usually represented by multiple voltage levels or time durations, depending on the required

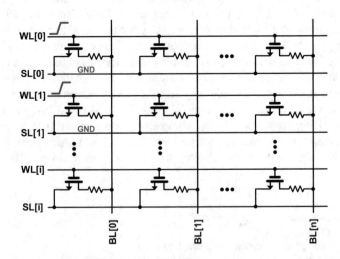

Fig. 5.1 ReRAM array for PIM

number of bits. W_i is stored in ReRAM devices and also requires multiple cells for realizing a multi-bit weight. The multiple cells can be located at multiple rows in the same column or at multiple columns in the same row, which is determined by the employed ReRAM macro architecture. When one column is used for MAC operation, the bitline current of each column (BL[i]) represents a MAC result and is digitized by an analog-to-digital converter (ADC). Using multiple columns needs additional control circuits for merging the bitline currents from the multiple columns and generating the final output current. Digitization will be performed using the final output current. Recently, multi-bit ReRAM devices have been reported, which allows a ReRAM cell to store a multi-bit weight [18–21]. The number of ReRAM cells for generating a MAC result will decrease when employing the multi-bit ReRAM devices. Therefore, higher density ReRAM PIM macros can be realized using the same ReRAM array density. However, multi-bit ReRAM technology is not mature and shows large variations. Therefore, the application of it is still limited.

5.3   Multiplication in ReRAM PIMs

5.3.1   Binary Multiply

Table 5.1 shows how a ReRAM cell can be used for binary multiply. When two binary bits, input, and weight, are multiplied, the output will be either logic "1" or logic "0." In 1T1R ReRAM cell, the input is applied to the wordline (WL[i] in Fig. 5.1), and the weight is stored in the ReRAM device. In general, the high resistance state (HRS) and the low resistance state (LRS) represent "0" and "1," respectively. When the input is "0," the access transistor in the 1T1R cell is off, and no current will flow through the cell ($I_{MC} = 0$). When the input is "1," the access transistor is on, and the current will be determined by the ReRAM device state. If the ReRAM device in HRS (Weight = "0"), the current (I_{MC}) will be I_{HRS}. When the ReRAM device is in LRC, I_{LRS} will flow through the cell. Here, it needs to be noted that the binary multiply result of "0" is represented by two different current values (i.e., 0 and I_{HRS}), which will degrade the sensing margin. The impact of I_{HRS} will be more significant when a large number of rows are activated for the multiply-and-accumulate (MAC) operation. In this case, I_{HRS} in multiple ReRAM devices will be added. It becomes more difficult to generate accurate MAC results because of the unwanted I_{HRS}. Therefore, the number of rows to be accessed at the same time should be decided carefully after considering the impact of I_{HRS}. The impact of I_{HRS} will be mitigated by increasing the ratio of I_{LRS} to I_{HRS}.

Table 5.1 Binary multiply in ReRAM

Input (IN)	Weight (W)	Product (IN × W)	I_{MC}
0	0 (HRS)	0	0
0	1 (LRS)	0	0
1	1 (LRS)	1	I_{LRS}
1	0 (HRS)	0	I_{HRS}

Table 5.2 Multiply with ternary weight in ReRAM [11]

Input (IN)	Ternary weight	nvCIM-P			nvCIM-N		
		Weight (W)	IN × W	I_{MC}	Weight (W)	IN × W	I_{MC}
0	+1	+1 (LRS)	0	0	0 (HRS)	0	0
1		+1 (LRS)	+1	I_{LRS}	0 (HRS)	0	I_{HRS}
0	0	0 (HRS)	0	0	0 (HRS)	0	0
1		0 (HRS)	0	I_{HRS}	0 (HRS)	0	I_{HRS}
0	−1	0 (HRS)	0	0	−1 (LRS)	0	0
1		0 (HRS)	0	I_{HRS}	−1 (LRS)	−1	I_{LRS}

5.3.2 Multiplication with Ternary Weight

Multi-bit weights can be realized by using multiple ReRAM cells. Table 5.2 shows an example of realizing multiplication of 1-bit input (IN) with a ternary weight [11]. The ternary weight is implemented with two ReRAM cells, one in the positive array (nvCIM-P) and the other in the negative array (nvCIM-N). The weight in the positive array includes only "+1(LRS)" and "0(HRS)" while that in the negative array includes "-1(LRS)" and "0(HRS)." The ternary multiply result is obtained by combining the results from the positive and negative arrays as depicted in Table 5.2. The multiply result will be "1" only when the input and the weight in the positive array are "1" and "1(LRS)" and the weight in the negative array is "0(HRS)." The multiply result of "−1" is defined by the opposite case where the input and the weight are "1" and "−1(LRS)" in the negative array and the weight in the positive array is "0(HRS)." Similar to Table 5.1, various cases generate I_{HRS} even though their multiply results must be "0." The impact of I_{HRS} needs to be carefully handled to meet the required output precision.

5.3.3 Multi-bit Multiplication

5.3.3.1 Multiplication Using One Cycle and One Column

Multiplication of multi-bit inputs and multi-bit weights can be done in various ways. Figure 5.2 illustrates three different ways for realizing 2-bit input (IN[1:0]) and 2-bit weight ($W_M W_L$) multiplication. Figure 5.2a, b use one cycle, while Fig. 5.2c uses two cycles. However, Fig. 5.2a generates the multiplication result using one column while Fig. 5.2b, c use multiple columns for generating a MAC result. In Fig. 5.2a, the 2-bit weight for IN[0] (LSB) can be realized by using three cells, one cell for W_L (LSB) and two cells for W_M (MSB). However, for IN[1] (MSB), six cells are necessary since IN[1] is 2 × IN[0] when both are "1s." Therefore, total nine ReRAM cells are necessary to realize the multiplication of the 2-bit input and the 2-bit weight. This will produce the maximum current of $9I_{cell}$. Even though Fig. 5.2a can realize the multi-bit multiplication using one column in principle, it is challenging to generate large bitline current accurately. One of the main reasons is that the range of the bitline voltage during read operation is limited by the ReRAM device characteristics. If the bitline voltage is relatively high, the accessed ReRAM devices are under weak set or reset conditions, which can partially change the ReRAM resistance. To avoid this, the bitline voltage should be maintained low so that no disturbance occurs in the ReRAM resistance. However, when the bitline voltage is maintained low, the current precision will also be affected, which limits the overall multiplication accuracy. Besides the low bitline voltage, ReRAM device variations also make it challenging to generate accurate bitline current proportional to the MAC result.

5.3.3.2 Parallel-Input Parallel-Weight (PIPW)

To improve the multiplication accuracy, we can use multiple columns, multiple macros, or multiple cycles assisted with extra circuits for merging the split currents from multiple columns, multiple macros, or multiple cycles. In Fig. 5.2b, the LSB and the MSB are stored in Macro[0] and Macro[1], respectively. Two macros store the same weight, while IN[0] and IN[1] are applied to Macro[0] and Macro[1], respectively. The MSB of the weight is implemented with two ReRAM devices using two columns. The maximum bitline current is 2 × I_{cell} from the column storing MSB of the weight. Similarly, the column storing LSB of the weight produces maximum current of I_{cell}. The currents from two macros need to be merged to generate a multiplication result. When merging, the weight difference between the two macros should be considered. This can be done either by multiplying the current from the macro storing MSB by 2 or dividing the current from the macro storing LSB by 2. This architecture is called 'Parallel-Input-Parallel-Weight (PIPW)' [18].

5.3.3.3 Serial-Input Parallel-Weight (SIPW)

Another way of improving the multiplication accuracy is to use one macro over multiple cycles, which is called "Serial-Input-Parallel-Weight (SIPW)." As shown in Fig. 5.2c, the 2-bit input signals are applied to the macro bit by bit over two cycles. The maximum bitline current is 3 × I_{cell} in each cycle. To merge the currents over two cycles, the current from the first cycle needs to be sampled in a capacitor so that it can be merged with the current generated in the second cycle. In addition, the

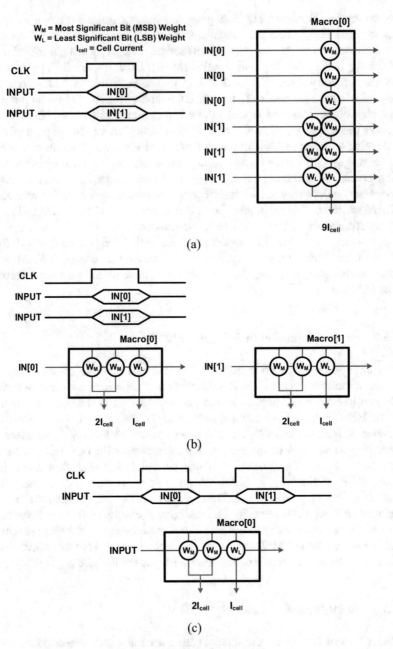

Fig. 5.2 Multiplication of multi-bit inputs and weights: (**a**) one cycle and one macro, (**b**) one cycle and multiple macros, and (**c**) multiple cycles and one macro [19]

Table 5.3 Ways of input weight multiplication

	Parallel-input/parallel-weight (PIPW)	Serial-input/parallel-weight (SIPW)
Input (m-bit)	WL cycle: $1\times$	WL cycle: $m\times$
Weight (n-bit)	# of cells: $n\times$	# of cells: $1\times$
Challenges	Large BL current and large parasitic capacitance	

ratio of the current produced by the MSB to the current produced by LSB needs to be considered before merging. This architecture is called "Serial-Input-Parallel-Weight (SIPW)."

The MAC result from PIPW and SIPW can be written as follows:

$$\text{MAC} = \text{IN}[0] \times [2 \times W_M + W_L] + \text{IN}[1][2 \times W_M + W_L]$$

Table 5.3 summarizes the comparison of PIPW and SIPW [19]. The number of ReRAM cells can be reduced when multi-bit ReRAM devices are utilized. The multi-bit ReRAM devices can employ any multiplication methods in Fig. 5.2. However, it is challenging to implement multi-bit ReRAM devices accurately. Therefore, the output precision of the multi-bit ReRAM-based multiplication is still limited.

5.4 ReRAM PIM Architecture

5.4.1 Introduction

This section will introduce the overview of ReRAM PIM architecture. Convolutional neural networks (CNNs) have demonstrated high accuracy in various artificial intelligence (AI) tasks. CNNs consist of multiple convolutional layers and fully connected layers as shown in Fig. 5.3a. Dot product and multiply-and-accumulate (MAC) operations are the basic operations that are heavily executed in CNNs as illustrated in Fig. 5.3b. These functions consume excessive energy in Von Neumann architecture because of the massive data transfer between memory and processing elements. Processing-in-memory (PIM) can address this issue by merging memory and processing elements together. However, typical PIMs utilize analog signals to represent MAC results, which requires careful design, particularly when high output precision is necessary. As explained in Fig. 5.2, the summation of the multiplication results can be implemented in various ways depending on the accuracy of the current in each bitline. The activation function (f) in Fig. 5.3b is generally realized by an analog-to-digital converter (ADC) for multi-bit precision or a comparator for one-bit precision.

Typical deep neural networks (DNNs) store weights in separated memories in non-volatile memories and transfer them to processing elements through multiple

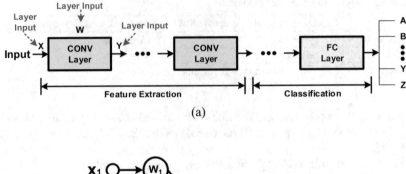

(a)

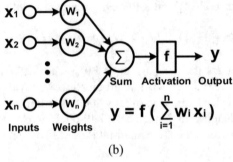

$$y = f\left(\sum_{i=1}^{n} w_i x_i\right)$$

(b)

Fig. 5.3 (a) CNN structure and (b) multiply-and-accumulate operation [22]

memory layers such as DRAM and SRAM. While processing-in-memory can reduce the amount of weight transfer significantly, weight transfer from memory to processing elements is still necessary. The weight transfer is particularly critical in edge computing for smart IoT devices where systems mostly stay in standby mode. Therefore, it is essential to minimize the standby power of DNNs. ReRAM-based PIM can minimize the standby power by disabling the supply voltage of the PIM blocks without losing the weight.

5.4.2 Non-volatile PIM Processor

Figure 5.4a depicts an example of non-volatile PIM-based processor architecture. Multiple non-volatile PIM (nvPIM) blocks store weights even without power supply, which allows the removal of external memory for storing weights when the processor is in standby operation. This can further reduce the data transfer between memory and processing elements leading to additional reduction in power and energy consumption. When compared to the conventional Von Neumann architecture with external non-volatile memory, the nvPIM can improve the energy efficiency 10–1000 times depending on the DNN architecture, input, weight, and output precision (Fig. 5.4b).

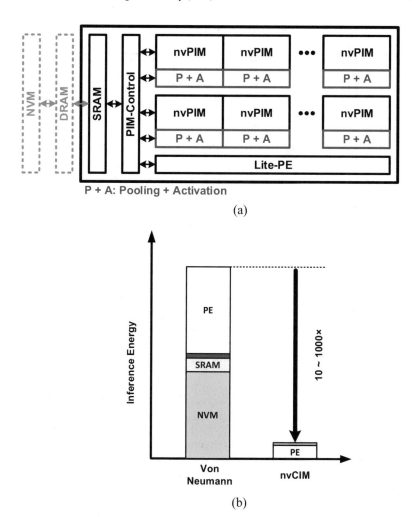

P + A: Pooling + Activation

(a)

(b)

Fig. 5.4 (**a**) PIM processor architecture and (**b**) inference energy comparison [11]

5.4.3 ReRAM PIM Architecture

Figure 5.5 shows a typical ReRAM PIM architecture. It consists of a ReRAM array, a row decoder, a reference generator, analog-to-digital converters (ADCs), and a write control block. The row decoder controls the input signals applied to the ReRAM array for PIM operation. Once bitline current is produced by the multiplication of input signal and weight, ADCs convert the bitline current into digital output for further processing in neural networks. The reference generator sets the conversion range of the ADCs, which varies depending on PVT variations. For better output precision, it is required for the reference generator to adjust

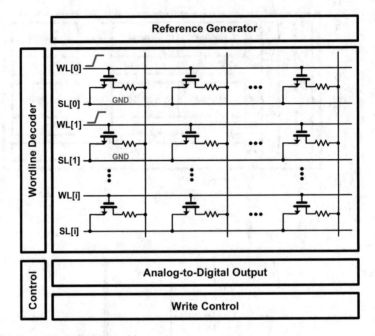

Fig. 5.5 Typical ReRAM PIM architecture

the conversion range automatically after tracking the actual variations. In general, reference voltage levels are generated by using ReRAM replicas for tracking the systematic variations. However, ReRAM devices show large variations compared to CMOS transistors. Therefore, the output precision of ReRAM-based PIMs is still worse than that of CMOS counterparts. Mature ReRAM technology with less device-to-device mismatches will reduce the precision gap between CMOS-based PIMs and ReRAM-based PIMs.

5.4.4 ADCs and DACs in ReRAM PIM

Figure 5.6 illustrates the traditional non-volatile ReRAM-based PIM architecture and a sample power and area breakdown of a ReRAM PIM macro in [14]. The ReRAM array size is 1152×128. Two-bit DACs and 8-bit ADCs are considered in the breakdown evaluation. Unlike normal ReRAM where only one row is activated at a time, and sense amplifiers produce binary comparison results, ReRAM PIM requires digital-to-analog converters (DACs) for input data and analog-to-digital converters (ADCs) for MAC output. It is noticeable that the power of ADCs and DACs is dominant compared to that of a ReRAM array. The ADCs also occupy majority of the area when the required output precision is relatively high (e.g., 8 bit). For reducing the power consumption of the ADCs, power/energy-efficient

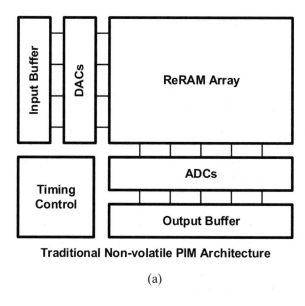

Traditional Non-volatile PIM Architecture

(a)

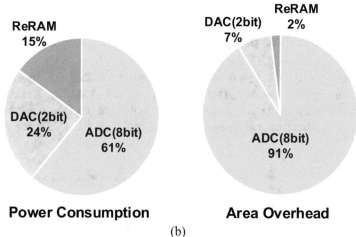

(b)

Fig. 5.6 (a) Traditional ReRAM-based PIM architecture and (b) sample power and area break-down [14]

successive-approximation-register (SAR) ADCs are widely employed. Besides, the number of ADCs can be reduced by reducing the number of MAC results that are generated at the same time. However, this will increase the number of cycles for processing the same amount of MAC results, degrading the performance of the PIMs. Even though various ADC techniques such as zero-skipping and column ADCs have been developed for better power and area efficiency in the SRAM-based PIMs [23–25], the power and area overheads of the ADCs are still challenging in the ReRAM-based PIMs.

5.5 ReRAM Co-processor

5.5.1 Architecture

A ReRAM crossbar is a compact solution for realizing vector-matrix multiplication by using the conductance values of the crossbar array storing the weights. By carefully controlling the voltage input for the crossbar rows, each column can produce current generated by the vector-matrix multiplication in the analog domain. It is well known that the relationship between the applied voltage and the induced current is not linear in the ReRAM crossbar. Therefore, pulse-width modulation is more commonly employed for applying the input signal to the ReRAM crossbar. Figure 5.7 illustrates the architecture of a fully integrated reprogrammable crossbar ReRAM coprocessor [26]. It consists of a RIC CPU, multiple SRAMs, a memory controller, a ReRAM PIM macro with mixed-signal interface. The CPU controls the ReRAM PIM macro and the mixed-signal interface through the shared bus. The ADCs and the DACs are controlled through the global configuration register depending on the operation mode of the coprocessor.

5.5.2 Mixed-Signal Interface

The block diagram of the mixed-signal interface of the ReRAM coprocessor is illustrated in Fig. 5.8. The ADC/DAC enables control block, and the 3b DAC select block determines the operations modes of the ADCs and the DACs. For example, during programming operation, the DACs for rows and the DACs for columns

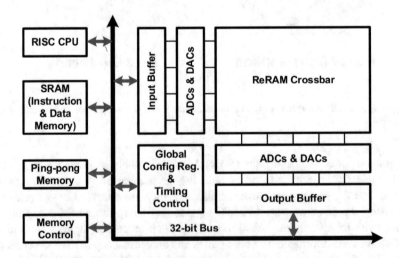

Fig. 5.7 Architecture of ReRAM coprocessor [26]

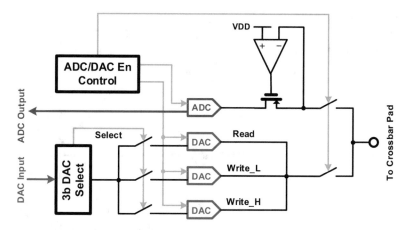

Fig. 5.8 Mixed-signal interface of ReRAM coprocessor [26]

generate high-voltage pulses to the accessed ReRAM cells for programming. During PIM operation, the DACs for read apply input pulses to the crossbar, and each column generates bitline current or voltage as a vector-matrix multiplication result. The ADCs connected to the selected columns convert the analog signal in each column into digital codes for further processing. The unused ADCs and DACs are disabled to reduce unnecessary power consumption.

5.5.3   ADCs and DACs Operation

Figure 5.9 depicts the operations of the ADCs and DACs during the programming and PIM modes of the ReRAM PIM coprocessor [26]. During programming (Fig. 5.9a), the DACs in the selected row and the selected column generate a train of differential pulses so that the selected ReRAM device undergoes either positive or negative programming voltage depending on the programming data. The DAC output voltage levels and the number of pulses are programmed into registers for flexible control. The DACs in the unselected rows and columns generate common-mode voltage (e.g., 1 V in [26]) to avoid unwanted programming. The driving strength of the DACs should be designed carefully so that the sneak currents flowing from the selected row to the unselected columns and from the unselected rows to the selected column do not affect the DAC outputs for the selected row and the selected column significantly. Since the worst-case scenarios need to be considered, the above requirement will increase the power and the area of the DACs. Figure 5.9b shows the ADCs and the DACs in the PIM mode where the DACs for rows are selectively activated relying on the applied input signal, and the ADCs for columns convert the conductance of each column into multi-bit digital outputs. The DACs for columns are disabled by the mixed-signal interface as shown in Fig. 5.8. The

Fig. 5.9 Operation of ADC
and DAC in ReRAM PIM:
(**a**) programming and (**b**) PIM
operation [26]

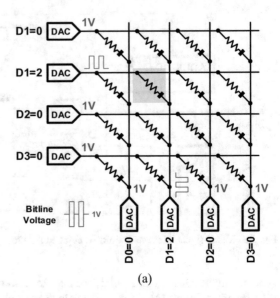

(a)

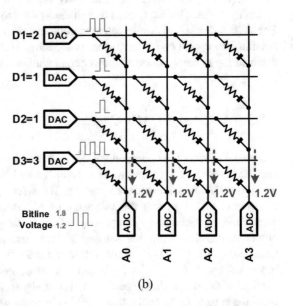

(b)

number of pulses going to each selected row is controlled by the controller (i.e.,
3b DAC Select in Fig. 5.8). The DACs for unselected rows generate 1.2 V as the
common-mode voltage. The DACs use the pulse amplitude of 0.6 V that can also be
adjusted depending on the ReRAM characteristics and the resolution of the ADCs.
The column outputs excited with 0.6 V pulses are digitized by the ADCs in parallel.

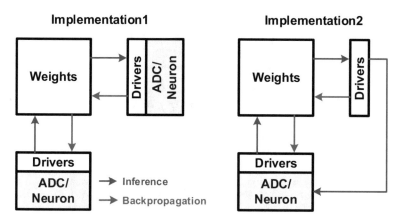

Fig. 5.10 Transposable ReRAM PIM macro [27]

5.6 Transposable ReRAM for Inference and Training

Transposable PIM macros with forward and backward propagation are necessary neural networks for inference and training [27]. In the inference mode, the forward operation will be performed. In the training mode, weights will be updated through backward propagation. Both propagation cases execute the convolution of weights and inputs. However, the weights matrix will be transposed in backward propagation compared with the feedforward one. Thus, the transpose weight matrix is necessary for computation. To realize transposable PIM macros, the memory array storing weights needs to be accessed vertically and horizontally. Figure 5.10 depicts two different architectures of the transposable PIM macros [27]. The first implementation includes dedicated drivers, ADCs, and neurons for forward propagation and backward propagation, respectively. However, the ADCs and the neurons can be used only one direction at a time, which facilitates sharing by the two propagation directions. The second implementation shows that the ADCs and the neurons are shared by the inference propagation and the backpropagation. This reduces the energy, latency, and area overheads coming from the ADCs and the neurons.

5.7 Bitline Sensing for MAC Accuracy Improvement

5.7.1 Variations in Bitline Current

ReRAM PIMs face various challenges such as large bitline current, large offset in sensing, overlap in bitline current for different MAC values, etc. Fig. 5.11a shows a ReRAM PIM macro activating multiple wordlines simultaneously. The bitline

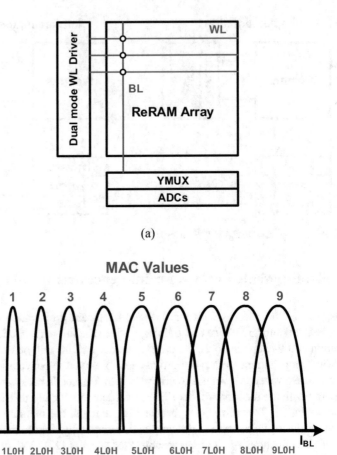

(a)

(b)

Fig. 5.11 (**a**) ReRAM PIM macro for parallel computation and (**b**) distribution of bitline current over multiple MAC values [11]

current distribution of each MAC value can be estimated depending on the ReRAM device status. Figure 5.11b illustrates an example of the bitline current distribution when assuming that the maximum MAC value from each column is 9 [11]. Here, the smallest bitline current for the MAC value of "1" is "1L0H" (one I_{LRS} and no I_{HRS}), which can happen when only one wordline is turned on for MAC operation. The maximum bitline current for the same MAC value is "1L8H" (one I_{LRS} and eight I_{HRS}) where 9 wordlines are turned on and 8 ReRAM devices are in the HRS state. The selected ReRAM devices in the HRS state will generate I_{HRS} even though the computed MAC value has no difference, which needs to be considered when sensing or digitizing the accumulated bitline current. Sensing margins can be defined by the current difference between two neighboring current distributions. For example, the sensing margin between the MAC value of "1" and that of "2" can be written as "2L0H" − "1L8H." Similarly, the sensing margin between the MAC value of "7" and that of "8" is "8L0H" − "7L2H." Ideally, the current difference between the MAC value of "7" and that of "8" (i.e. $I_{LRS} - 2I_{HRS}$) is larger than the current difference between the MAC value of "1" and that of "2" (i.e., $I_{LRS} - 8I_{HRS}$). However, after considering the variations in I_{LRS} and I_{HRS}, it is found that the current for higher MAC values shows larger variations, which degrades the sensing margins as depicted in Fig. 5.11b.

5.7.2   Input-Aware Dynamic Reference Generation

To tackle the sensing margin degradation issue caused by the ReRAM current variations, input-aware dynamic reference generation scheme is proposed in [11]. This scheme considers the reference current dependency on the input signal. Therefore, instead of using fixed reference currents, the reference currents are dynamically generated by input-aware replica rows. The replica rows are controlled by the number of wordlines for the input signal, which is counted by a counter. Figure 5.12 illustrates the distributions of the bitline current over various input values. It is evident that the optimal reference current for sensing MAC values reply on the number of wordlines (NWL). The input-aware reference current generation

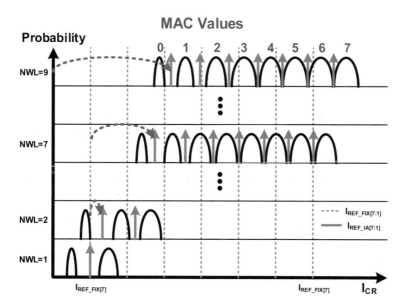

Fig. 5.12 Input-aware dynamic reference generation scheme [11]

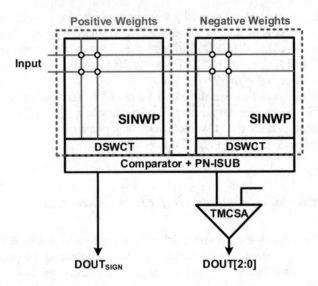

Fig. 5.13 PIM macro architecture of SINWP [18]

is more critical for higher MAC values where the current difference between MAC values is smaller because of the variations in the bitline current. It is reported that the input-aware dynamic reference generation improves the error rate by 50 times compared to the conventional fixed reference generation scheme.

5.7.3 Weighted Current Generation

5.7.3.1 PIM Macro Architecture

Generally, multi-bit weights are stored in multiple ReRAM cells to generate weighted currents. This increases area overheads and bitline current. Since the bitline current shows significant variations because of the ReRAM resistance variations, using large bitline current for MAC operation limits the output precision. To address the area and the bitline current issues, a research work proposing serial-input non-weighted product (SINWP) and down-scaling weighted current translation (DSWCT) is reported in [18]. Figure 5.13 shows the overall array architecture of the PIM macro in [18]. The macro consists of two ReRAM arrays for positive weights and negative weights, respectively. The current from each array will be weighted through the DSWCT block and will be merged by the positive/negative current subtraction (PN-ISUB) block.

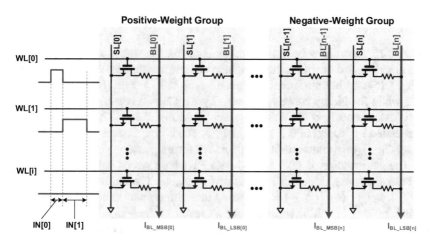

Fig. 5.14 ReRAM PIM macro operation for SINWP [18]

Table 5.4 Realization of positive and negative weights in SINWP

Positive weight group			Negative weight group		
MC_M	MC_L	Weight (W)	MC_M	MC_L	Weight (W)
LRS (+2)	LRS (+1)	+3	HRS (0)	HRS (0)	0
LRS (+2)	HRS (0)	+2	HRS (0)	LRS (−1)	−1
HRS (0)	LRS (+1)	+1	LRS (−2)	HRS (0)	−2
HRS (0)	HRS (0)	0	LRS (−2)	LRS (−1)	−3

5.7.3.2   Serial-Input Non-weighted Product (SINWP)

Figure 5.14 shows the schematic of the ReRAM array for SINWP. The 2-bit input is applied to the array at different timings. LSB is applied first, while MSB is applied later. The array stores 2-bit weights using two ReRAM bit cells that are implemented in two columns. The bitline currents from two columns are non-weighted. Therefore, they need to be further processed to generate final weighted current for digitization. The array for the negative weights also generates bitline currents in the same way as the array for the positive weights.

Table 5.4 explains the realization of the positive and negative weights in SINWP. Note that one resistance combination of two ReRAM bit cells indicates two weight values with the same absolute value. The combination of (MC_M = LRS, MC_L = LRS) is used for the weights of "3" and "−3" since it produces the largest bitline current. Similarly, the weights of "1" and "−1" are realized by the combination of (MC_M = HRS, MC_L = HRS) for generating the smallest bitline current.

5.7.3.3 Down-scaling Weighted Current Translator (DSWCT)

Figure 5.15 shows how two bitline currents from each weight group are weighted by a circuit called down-scaling weighted current translator (DSWCT). As shown in Fig. 5.15a, the bitline current generated by the MSB weight (I_{MSB}) is downscaled by 2, while the bitline current generated by the LSB weight (I_{LSB}) is downscaled by 4. The schematic of DSWCT is presented in Fig. 5.15b. The weighted current is generated by $N0$, $N1$, $P0$, $P1$, $P2$, and $P3$. The bitline current generated by MSB weight ($I_{DL\_MSB[0]}$) flows through $P0$ and $N0$ forming analog voltage at the gate of $P0$. The gate voltage of $P0$ is shared with $P1$ whose size is half of $P0$. Therefore the current flowing through $P1$ is half of $I_{DL\_MSB[0]}$. The bitline current generated by LSB weight ($I_{DL\_LSB[0]}$) is processed in a similar way except that the size of $P3$ is a quarter of $P2$. Therefore, only a quarter of $I_{DL\_LSB[0]}$ flows through $P3$. As shown in Fig. 5.14, $I_{DL\_MSB[0]}$ and $I_{DL\_LSB[0]}$ are generated when IN[0] is

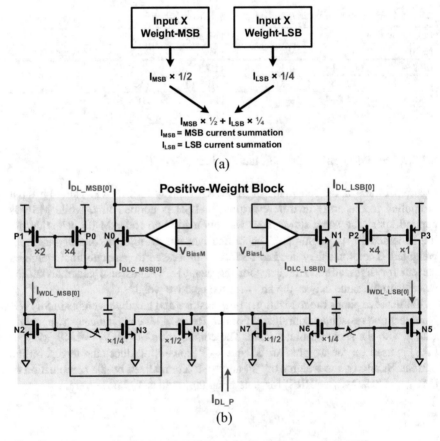

Fig. 5.15 (**a**) Operation principle of DSWCT and (**b**) circuit implementation [18]

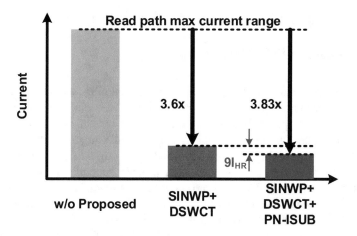

Fig. 5.16 Read path current reduction by SINWP+DSWCT+PN-ISUB [18]

applied to the ReRAM array. These two current components need to be merged with the current generated by IN[1]. Therefore, $I_{\text{WDL\_MSB}[0]}$ and $I_{\text{WDL\_LSB}[0]}$ are stored in the capacitors after converting them into voltage at the gate node of $N2$ and $N5$. After this, IN[1] is applied to the array and generates bitline currents, $I_{\text{DL\_MSB}[1]}$ and $I_{\text{DL\_LSB}[1]}$. Note that these currents are not weighted when compared with $I_{\text{DL\_MSB}[0]}$ and $I_{\text{DL\_LSB}[0]}$. $N3$, $N4$, $N6$, and $N7$ merge $I_{\text{DL\_MSB}[0]}$, $I_{\text{DL\_LSB}[0]}$, $I_{\text{DL\_MSB}[1]}$, and $I_{\text{DL\_LSB}[1]}$ after considering their weights. Since the weight of IN[1] is $2\times$ of IN[0], $I_{\text{WDL\_MSB}[0]}$ and $I_{\text{WDL\_LSB}[0]}$ are downscaled by 4 and $I_{\text{DL\_MSB}[1]}$ and $I_{\text{DL\_LSB}[1]}$ are downscaled by 2 for proper merging. This is realized by selecting the device size of $N3$ and $N6$ a quarter of $N2$ and $N5$. Consequently, the merged current ($I_{\text{DL\_P}}$) can be written as follows.

$$I_{\text{DL\_P}} = \frac{1}{4}\left[I_{\text{WDL\_LSB}[0]} + I_{\text{WDL\_MSB}[0]}\right] + \frac{1}{2}\left[I_{\text{WDL\_LSB}[1]} + I_{\text{WDL\_MSB}[1]}\right]$$

$$= \frac{1}{4}\left[\frac{I_{DL\_\text{LSB}[0]}}{4} + \frac{I_{DL\_\text{MSB}[0]}}{2}\right] + \frac{1}{2}\left[\frac{I_{DL\_\text{LSB}[1]}}{4} + \frac{I_{DL\_\text{MSB}[1]}}{2}\right]$$

The merged current from the negative weight group ($I_{\text{DL\_N}}$) is also generated by the same way as $I_{\text{DL\_P}}$. Finally, "$I_{\text{DL\_P}} - I_{\text{DL\_N}}$" is computed by PN-ISUB whose output is digitized. Figure 5.16 shows the read path current reduction achieved by SINWP, DSWCT, and PN-ISUB. The improvement of $3.83\times$ is obtained.

5.8 Versatile ReRAM-Based PIM Functions

5.8.1 Versatile PIM Architecture

In data-intensive applications such as machine learning and neural networks, Processing-in-Memory (PIM) is an attractive way of reducing energy and latency. Even though Multiply-and-Accumulate (MAC) is one of the most commonly executed functions in neural networks, modern SoCs require other data-centric functions such as logic functions, addition, and other memory operations as well. For example, the ReRAM macros in [28–30] support both ternary content addressable memory (TCAM) operation as well as regular memory operation. Various ReRAM-based cell structures such as 4T2R and 2T2R are reported to support various PIM operation modes [15, 16]. Figure 5.17 illustrates an array architecture of 2T2R ReRAM for versatile functions [15]. It consists of an array, various decoders, drivers, reconfigurable sense amplifiers (SAs), and PIM logic. This ReRAM macro supports various functions such as TCAM, in-memory dot product, logic-in-memory, and normal ReRAM operation. This will be useful in applications where multiple memory functions are required without using dedicated memory for each function.

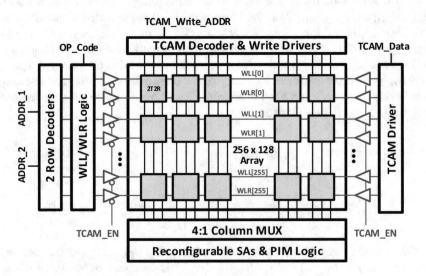

Fig. 5.17 2T2R ReRAM for versatile PIM functions [15]

5.8.2 2T2R ReRAM Bit Cell for Versatile Functions

5.8.2.1 Basic Memory Operation

Figure 5.18 shows the 2T2R bit cell and its normal ReRAM operation employed in Fig. 5.17. It comprises two 1T1R ReRAM bit cells using a common source line (SL) scheme [14, 31] and stores differential data. For read operation (Fig. 5.18b), both WLL and WLR are enabled so that SL and SLB generate differential voltage that can be sensed by a sense amplifier. The 2T2R bit cell employs a two-cycle write operation. During the first cycle, BL is set to the set voltage (V_{SET}), and SL and SLB are grounded. This will set the ReRAM devices to LRS. At the second cycle, one of SL and SLB is set to the set voltage (V_{SET}) while the unselected SL or SLB, and BL are grounded. This will make the selected ReRAM device transit to HRS. Therefore, differential data can be written into the 2T2R bit cell over two cycles. This may not be preferred in applications where the frequent write operation is executed. However, in PIMs, read operation in various modes is much more frequent. Therefore, it is acceptable to sacrifice one cycle for the differential data writing.

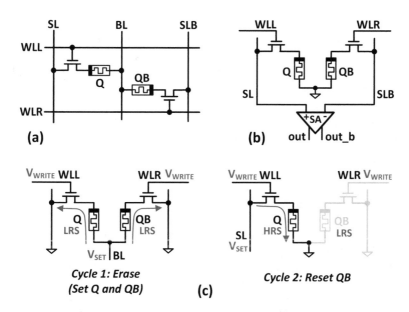

Fig. 5.18 2T2R ReRAM bitcell: (**a**) schematic, (**b**) read, and (**c**) write [15]

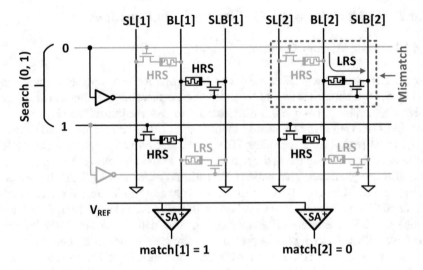

Fig. 5.19 2T2R ReRAM operation in the TCAM mode [15]

5.8.2.2 TCAM Operation

Figure 5.19 explains the 2T2R ReRAM operation in the TCAM mode. Search data are loaded into WLL (e.g., (0,1) in Fig. 5.18), and the bitlines (BL[i]) and the source lines (SL[i] and SLB[i]) are precharged to VDD and grounded, respectively. When there is a mismatch, the corresponding bitline is discharged quickly through the mismatched cell and the sense amplifier in the corresponding bitline will produce "0" as a result. If the number of mismatched cells increases, the discharging speed will be higher. The overall search operation result will be generated through the sense amplifiers.

5.8.2.3 Logic-in-Memory Operation

The 2T2R ReRAM in Fig. 5.17 also supports logic-in-memory operation as shown in Fig. 5.20. AND/NAND operations are executed by enabling two WLLs in two rows with grounded BLs. Similarly, OR/NOR operations are performed by enabling two WLRs in two rows with grounded BLs. XOR operation (Fig. 5.20a) can be done by combining the results of the AND and NOR operations. XNOR operation (Fig. 5.20b) can also be realized in a similar way with grounded SLs. Here, BLs are connected to the sense amplifiers through reconfiguration. Figure 5.20c shows the logic functions required for a full adder (FA) and a full subtractor (FS). Note that all the logic functions can be achieved by the circuit configurations explained in Fig. 5.20a, b. However, FA requires two cycles since the sense amplifier output from Fig. 5.20a needs to be latched before using the configuration of Fig. 5.20b.

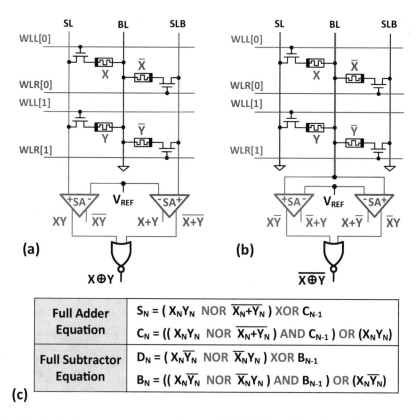

Fig. 5.20 Logic-in-memory operation: (**a**) grounded SL/SLB and BL, (**b**) precharged BL and grounded SL/SLB, and (**c**) full adder and full subtractor equations [15]

5.8.2.4   Dot Product Operation

In binary neural networks (BNNs), doc product operation is realized by simple XNOR-popcount operation. Various research works for implementing in-memory dot product (IM-DP) operation have been reported [11, 32]. Figure 5.21 illustrates the IM-DP operation proposed in Fig. 5.17. The ReRAM array stores weights using a 2T2R bit cell, and the input activation signal (F) is applied to the pass gates for controlling the connectivity of SLs for the sense amplifiers. When $F = $ "1," SL and SLB are connected to the positive input and the negative input, respectively. When $F = $ "0," SL and SLB are connected to the opposite sense amplifier inputs. When one row is enabled by turning on the corresponding WLL and WLR, the sense amplifier outputs will be the bitwise XOR/XNOR of the selected weight (W) and the input activation signal (F). The sense amplifier outputs go through a Wallace tree adder to execute the popcount operation. Since the popcount operation is realized in a fully digital manner, no analog-to-digital converters (ADCs) are necessary, which

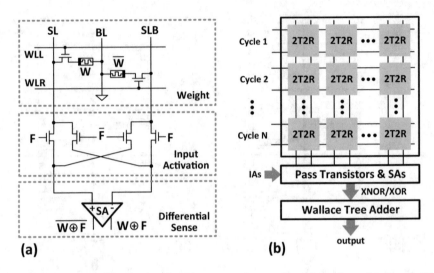

Fig. 5.21 In-memory dot product for binary neural networks: (**a**) sensing scheme and (**b**) simplified architecture [15]

Table 5.5 Comparison with recent ReRAM and R-CIM works

		This work	[15]	[30]	[31]
Operations		NVM, TCAM, LiM, IM-DP	NVM, IM-DP	TCAM	NVM
Technology		40 nm	130 nm	130 nm	40 nm
On/Off ratio		~100	N.A.	> 200	N.A.
V_{READ} (V)	TCAM	0.4	–	0.6	–
	LiM	0.15	–	–	–
	IM-DP	0.1	0.1	–	–
	NVM read	0.1	0.1	–	0.18–0.3

reduces power and area overheads significantly. Table 5.5 compares various ReRAM supporting various PIM functions without MAC operation.

5.9 Summary

This chapter presents an overview of ReRAM PIMs. Non-volatile ReRAM PIMs are attractive for edge devices whose power is from a battery or energy harvesting devices. Even though many techniques developed for SRAM-based PIMs and DRAM-based PIMs can be considered for ReRAM PIMs, ReRAM device characteristics incur various design challenges. One of the most critical challenges in the ReRAM PIM design is to tackle large variations in the ReRAM device characteristics. It is also a limiting factor that the bitline voltage and the bitline current for PIM operation should be lowered as much as possible to minimize the nonlinearity. This chapter introduces various design techniques, including cell-level techniques, ReRAM PIM architectures, and CMOS circuit techniques for addressing the aforementioned challenges. ReRAM PIMs will be more impactful when the ReRAM fabrication technology becomes more mature and the key ReRAM device parameters are improved.

References

1. K. Ando et al., BRein memory: A single-chip binary/ternary reconfigurable in-memory deep neural network accelerator achieving 1.4 TOPS at 0.6 W. IEEE J. Solid State Circuits 53(4), 983–994 (2018)
2. M. Kang et al., A multi-functional in-memory inference processor using a standard 6T SRAM array. IEEE J. Solid State Circuits 53(2), 642–655 (2018)
3. A. Biswas et al., CONV-SRAM: An energy-efficient SRAM with in-memory dot-product computation for low-power convolutional neural networks. IEEE J. Solid State Circuits 54(1), 217–230 (2019)
4. H. Valavi et al., A 64-tile 2.4-Mb in-memory-computing CNN accelerator employing charge-domain compute. IEEE J. Solid State Circuits 54(6), 1789–1799 (2019)
5. S. Yin et al., XNOR-SRAM: In-memory computing SRAM macro for binary/ternary deep neural networks. IEEE J. Solid State Circuits 55(6), 1733–1743 (2020)
6. S.K. Gonugondla et al., A 42pJ/decision 3.12TOPS/W robust in-memory machine learning classifier with on-chip training, in *Proc. IEEE int. solid-state circuits conf. (ISSCC)*, (IEEE, Piscataway, 2018), pp. 490–492
7. J. Su et al., A 28nm 64Kb inference-training two-way transpose multibit 6T SRAM compute-in-memory macro for AI edge chips, in *Proc. IEEE int. solid-state circuits conf. (ISSCC)*, (IEEE, Piscataway, 2020), pp. 240–242
8. X. Si et al., A 28nm 65Kb 6T SRAM computing-in-memory macro with 8b MAC operation for AI edge chips, in *Proc. IEEE int. solid-state circuits conf. (ISSCC)*, (IEEE, Piscataway, 2020), pp. 246–248
9. W. Khwa et al., A 65nm 4Kb algorithm-dependent computing-in-memory SRAM unit-macro with 2.3ns and 55.8TOPS/W fully parallel product-sum operation for binary DNN edge processors, in *Proc. IEEE int. solid-state circuits conf. (ISSCC)*, (IEEE, Piscataway, 2018), pp. 496–498
10. X. Si et al., A twin-8T SRAM computation-in-memory macro for multiple-bit CNN-based machine learning, in *Proc. IEEE int. solid-state circuits conf. (ISSCC)*, (IEEE, Piscataway, 2019), pp. 396–398
11. W.-H. Chen et al., A 65nm 1Mb non-volatile computing-in-memory ReRAM macro with sub-16ns multiply-and-accumulate for binary DNN AI edge processors, in *Proc. IEEE int. solid-state circuits conf. (ISSCC)*, (IEEE, Piscataway, 2018), pp. 494–496
12. C. Xue et al., Embedded 1-Mb ReRAM-based computing-in-memory macro with multibit input and weight for CNN-based AI edge processors. IEEE J. Solid State Circuits 55, 203–215 (2020)
13. C. Xue et al., A 22nm 2Mb ReRAM compute-in-memory macro with 121-28TOPS/W for multibit MAC computing for tiny AI edge devices, in *Proc. IEEE int. solid-state circuits conf. (ISSCC)*, (IEEE, Piscataway, 2020), pp. 244–246
14. Q. Liu et al., A fully integrated analog ReRAM based 78.4TOPS/W compute-in-memory chip with fully parallel MAC computing, in *Proc. IEEE int. solid-state circuits conf. (ISSCC)*, (IEEE, Piscataway, 2020), pp. 500–502
15. Y. Chen et al., Reconfigurable 2T2R ReRAM architecture for versatile data storage and computing in-memory. IEEE Trans. VLSI Syst. 28, 2636–2649 (2020)

16. Y. Chen et al., A reconfigurable 4T2R ReRAM computing in-memory macro for efficient edge applications. IEEE Open J. Circuits Syst. **2**, 210–222 (2021)
17. C.-X. Xue et al., A 22nm 4Mb 8b-precision ReRAM computing-in-memory macro with 11.91 to 195.7TOPS/W for tiny AI edge devices, in *Proc. IEEE int. solid- state circuits conf. (ISSCC)*, (IEEE, Piscataway, 2021), pp. 245–247
18. C.-X. Xue et al., A 1Mb multibit ReRAM computing-in-memory macro with 14.6ns parallel MAC computing time for CNN based AI edge processors, in *Proc. IEEE int. solid-state circuits conf. (ISSCC)*, (IEEE, Piscataway, 2019), pp. 388–390
19. W. Lee et al., Multilevel resistive-change memory operation of Al-doped ZnO thin-film transistor. IEEE Electron. Dev. Lett. **37**(8), 1014–1017 (2016)
20. R. Yasuhara et al., Reliability issues in analog ReRAM based neural-network processor, in *IEEE international reliability physics symposium (IRPS)*, (IEEE, Piscataway, 2019), pp. 1–5
21. R. Mochida et al., A 4M synapses integrated analog ReRAM based 66.5 TOPS/W neural-network processor with cell current controlled writing and flexible network architecture. IEEE Symp. VLSI Technol. **2018**, 175–176 (2018)
22. A. Biswas et al., Conv-RAM: An energy-efficient SRAM with embedded convolution computation for low-power CNN-based machine learning applications, in *Proc. IEEE int. solid- state circuits conf. (ISSCC)*, (IEEE, Piscataway, 2018), pp. 488–490
23. C. Yu et al., A 16K current-based 8T SRAM compute-in-memory macro with decoupled read/write and 1-5bit column ADC, in *Proc. IEEE custom integrated circuits conference (CICC)*, (2020)
24. C. Yu et al., A zero-skipping reconfigurable SRAM in-memory computing macro with binary-searching ADC, in *Proc. IEEE Eur. solid state circuits conf. (ESSCIRC)*, (2021)
25. C. Yu et al., A logic-compatible eDRAM compute-in-memory with embedded ADCs for processing neural networks. IEEE Trans. Circuits Syst. I Regul. Pap. **68**(2), 667–679 (2021)
26. J.M. Correll et al., A fully integrated reprogrammable CMOS-RRAM compute-in-memory coprocessor for neuromorphic applications. IEEE J. Explor. Solid-State Computat. Devices Circuits **6**(1), 36–44 (2020)
27. W. Wan et al., A 74 TMACS/W CMOS-RRAM neurosynaptic core with dynamically reconfigurable dataflow and in-situ transposable weights for probabilistic graphical models, in *Proc. IEEE intl. solid-state circuits conference (ISSCC)*, (2020), pp. 498–499
28. L. Zheng et al., Memristors-based ternary content addressable memory (mTCAM), in *IEEE int. symp. on circuits and systems (ISCAS)*, (2014), pp. 2253–2256
29. M. Chang et al., Designs of emerging memory based non-volatile TCAM for Internet-of-Things (IoT) and big-data processing: A 5T2R universal cell, in *IEEE int. symp. on circuits and systems (ISCAS)*, (IEEE, Piscataway, 2016), pp. 1142–1145
30. D. Ly et al., In-depth characterization of resistive memory-based ternary content addressable memories, in *IEEE int. electron devices meeting (IEDM)*, (IEEE, Piscataway, 2018), pp. 20.3.1–20.3.4
31. C. Chou et al., An N40 256K×44 embedded RRAM macro with SL-precharge SA and low-voltage current limiter to improve read and write performance, in *Proc. IEEE int. solid-state circuits conf. (ISSCC)*, (IEEE, Piscataway, 2018), pp. 478–479
32. M. Bocquet et al., In-memory and error-immune differential RRAM implementation of binarized deep neural networks, in *IEEE intl. electron devices meeting (IEDM)*, (IEEE, Piscataway, 2018), pp. 20.6.1–20.6.4

Chapter 6
PIM for ML Training

Jaehoon Heo and Joo-Young Kim

6.1 Introduction

Machine learning (ML) inference is the evaluation process of a trained model for a given input. To this end, it reads the input data and sends it through the various ML layers, such as fully connected, convolutional, and recurrent layers, which involve data-intensive computations with model parameters to get the final result. It is a read-only and unidirectional process. On the other hand, ML training is the process of finding the network's weight and bias parameters that can perform a target task. Mathematically speaking, it defines the cost function and updates the model parameters to minimize the cost for the given training data set consisting of many pairs of inputs and outputs. It involves numerous parameter updates with iterative forward and backward propagation.

With algorithmic complexity and limited usage, there are not many commercial products available or under development for ML training, except in the cloud datacenter domain. This is why GPU is still the most dominant platform in training, unlike in inference where many accelerators challenge to replace GPU. As discussed in the previous chapters, processing-in-memory (PIM) architecture can improve both performance and energy efficiency in various ML workloads by addressing the data movement bottleneck between the compute and memory device. Since the training process generates more intermediate data and requires higher bandwidth than the inference, PIM has greater opportunities in training, despite its computational complexity. In this chapter, we will review the training computations and look into the latest PIM works designed for ML training.

J. Heo · J.-Y. Kim (✉)
School of Electrical Engineering (E3-2), KAIST, Daejeon, South Korea
e-mail: kd01050@kaist.ac.kr; jooyoung1203@kaist.ac.kr

© The Author(s), under exclusive license to Springer Nature Switzerland AG 2023
J.-Y. Kim et al. (eds.), *Processing-in-Memory for AI*,
https://doi.org/10.1007/978-3-030-98781-7_6

6.2 Training Computations

Unlike the inference has a data-intensive but simple computational flow, the training process has a complex flow with many iterations. The goal of training is to find all the weight and bias parameters that minimize the cost function written in Eq. 6.1. It represents the overall distance between the predicted outputs y^o and the training sample outputs y_t.

$$argmin\left(C = \frac{1}{2}\|y^o - y_t\|^2\right) \tag{6.1}$$

Once the target model's weight and bias parameters are initialized, the training iterates the following steps for each training sample to minimize the cost function. First, it performs feed-forward propagation (FP). It is the same as the inference process; an input vector goes through the network for evaluation. It then computes the error at the output by subtracting the evaluation result and the training sample. Second, it propagates the error backward from the output to the input (backward propagation, BP). Third, it calculates the gradient for each layer to reduce the overall difference between the predicted outputs and the ground truths (i.e., training set). Lastly, it updates the weight and bias parameters. The above process is repeated until the network is converged, which means the magnitude of an update gets small enough under a threshold.

We use the gradient descent method for optimization, which iteratively moves in the steepest descent direction defined by the gradient's negative to minimize the cost function. With a stochastic process using mini-batching, stochastic gradient descent (SGD) is the de facto standard in training as it enables fast convergence. Equation 6.2 shows how the weights are updated in the SGD.

$$W^+ = W - \eta \cdot \frac{\partial E}{\partial W} \tag{6.2}$$

6.2.1 Feed-Forward Propagation

The input data is propagated through multiple layers in FP, executing multiply-and-accumulate (MAC) operations between the input and weight data for each layer, as discussed in Chap. 1. Eqs. 6.3 and 6.4 show the operation of the fully connected layer and convolutional layer, respectively, where function f is non-linear activation function. For convolutional layers, down-sampling pooling functions can be placed between layers.

$$\begin{cases} Y^l = W^l H^{l-1} + b \\ Z^l = f(Y^l) \end{cases} \tag{6.3}$$

$$\begin{cases} Y^l = W^l * H^{l-1} + b \\ Z^l = f(Y^l) \end{cases} \tag{6.4}$$

symbol $*$: convolution

Once FP is done, the final layer's output is the predicted result and is used for calculating the error against a labeled sample of the training set. Although the computation itself is the same as the inference, FP of the training process needs to keep the intermediate results because they will be re-used in the later steps. For each layer, it needs to store both activation and activation gradient, the gradient of the layer's activation function with respect to the layer's activation, as they are necessary for backward propagation and gradient calculation (GC). For the convolution layers with pooling, selected positions out of the pooling window are needed for error propagation during BP.

6.2.2 Backward Propagation

In BP, the calculated error δ is propagated layer by layer from the output to the input side. Based on the chain rule, we calculate the first-order derivative of the cost function with regard to each parameter to compute the propagated error and gradient matrix for each layer. Equation 6.5 shows the equation for fully connected layers, which multiplies the transposed weight matrix to the propagated error from the previous layer and applies the Hadamard product (i.e., element-wise multiplication) with the activation gradient. Likewise, Eq. 6.6 shows the equation for convolutional layers, which applies deconvolution instead of transposed multiplication. Deconvolution is same as the convolution with 180° rotated weights after zero padding.

$$\begin{cases} \delta^L = (Y - Y_t) \odot f^{L'}(Z^L) & \text{if Output layer L} \\ \delta^l = \left(\left(W^{l+1} \right)^T \delta^{l+1} \right) \odot f^{l'}(Z^l) & \text{if } l < L \end{cases} \tag{6.5}$$

$$\begin{cases} \delta^L = (Y - Y_t) \odot f^{L'}(Z^L) & \text{if Output layer L} \\ \delta^l = \left(W^{l+1} \star \delta^{l+1} \right) \odot f^{l'}(Z^l) & \text{if } l < L \end{cases} \tag{6.6}$$

symbol $\odot$: element-wise, $\star$: deconvolution

The error propagation process is different for average pooling and max pooling. The error is divided evenly with a square of the window size if it is average pooling. For the max pooling case, the error only propagates to the max positions stored during the FP stage.

6.2.3   Gradient Calculation and Weight Update

To calculate the gradient matrix, each layer operates the propagated error and the activation. As shown in Eqs. 6.7 and 6.8, outer product and convolution are applied for the fully connected and convolutional layer, respectively. Like in BP, GC utilizes the activation results stored during FP.

$$W^{l+} = W^l - \eta \cdot \delta^l \otimes H^{l-1} \tag{6.7}$$

$$W^{l+} = W^l - \eta \cdot \delta^l * H^{l-1} \tag{6.8}$$

symbol$\otimes$: outer-product, $*$: convolution

Finally, the weight parameters are updated by subtracting the multiplied product scaled by the learning rate. The learning rate is a hyperparameter that decides the magnitude of a moving step in SGD, deciding how fast the learning would be, while the gradient represents the direction of movement to minimize the loss. If the mini-batch size is more than one, the calculated gradient matrices should be averaged before the weight updated is performed.

6.3   SRAM-Based PIM for Training

Although SRAM-based PIM suffers from low memory density, its logic process is best to implement high-speed logic circuits. Some SRAM-based PIM works [1–3] have been proposed for on-device training, achieving high energy efficiency with good training accuracy.

6.3.1   Two-Way Transpose SRAM PIM

Su et al. [1] suggest the first SRAM-based PIM supporting both FP and BP stages of the training. Before this work, previous PIM chips mostly focused on low-energy inference scenarios for intelligent edge applications. This paper proposes a two-way transpose (TWT) SRAM macro that supports multi-bit MAC operations for FP and BP with high energy efficiency and compact area. It also contains customized sense amplifiers called small-offset gain-enhancement amplifiers to reduce energy consumption. The authors fabricated a chip in a 28 nm CMOS technology to verify the proposed design. Even though it is the first fabricated SRAM-based PIM that performs both FP and BP, it is impractical to be used in actual ML training because it does not cover the whole process, missing the GC and WU stage.

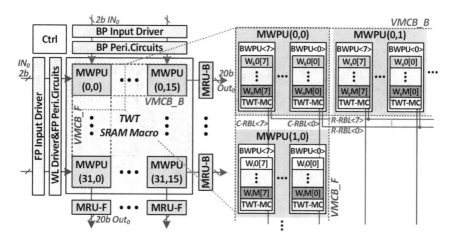

Fig. 6.1 Overall design of two-way transpose PIM

6.3.1.1 SRAM Compute-in-Memory Macro Design

Figure 6.1 shows the overall design of the TWT SRAM PIM macro. It consists of FP input driver, BP input driver, and 32×16 multibit-weight-product-units (MWPUs), whose total memory size is 64K bits. To support both multi-bit computation and digital conversion of the computed analog signal, the macro contains 16 multibit-readout units for FP (MRU-F) at the bottom and 32 MRUs for BP (MRU-B) on the right of the MWPU array. Each MWPU is composed of 8 bit-wise product units (BWPUs), in which each BWPU contains 16 SRAM cells in a column. An 8-bit weight is stored in the 8 cells on the same rows across the BWPUs. At the bottom of the BWPU, a TWT multiply cell (TWT-MC) exists to multiply the 1-bit weight from the cells and 2-bit input, utilizing voltage variation of bitline. The macro iterates multiple phases of 2-bit multiplication if the input bit width is greater than 2 and a multiple of 2.

6.3.1.2 In-memory Multiplication for Forward and Backward Propagation

Figure 6.2 illustrates how TWT macro performs in-memory multiplication in the BWPU. The TWT-MC inside BWPU has 2 pass transistors (N_1 and N_2) and 2×3 multiply transistors (N_3-N_5 and N_6-N_8). For the case of FP, the macro precharges column-read-bitline (C-RBL) to V_{DD} and sets row-read-bitline (R-RBL) to the ground. Then, it injects the two input bits, which is an activation data, via the forward wordline of MSB (FWLM) and LSB (FWLL). The FWLM and FWLL are connected to N_3 and N_7, respectively, and the weight is connected to both N_5 and N_6. Their values decide to either connect the path from the pre-charged bitline,

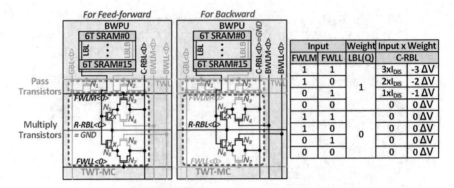

Fig. 6.2 BWPU Design and in-memory multiplication

C-RBL, to the ground or disconnect the path. If connected, the voltage drop occurs in C-RBL. To differentiate the amount of voltage drop according to the bit position of the input bits, the width of the multiply transistors whose gates are connected to a higher input bit, i.e., N_5 transistors, is double that of those connected to the lower input bit, i.e., N_6 transistors. This is because the current of a transistor increases with the gate width. As a result, the voltage drop that ranges from $3\Delta V$ to 0 is generated on C-RBL, and its value is equivalent to the multiplication between the 2-bit input and 1-bit weight, as delineated in the table. Since the 32 BWPUs across the MWPUs on the same column share the C-RBL, their voltage drops are all accumulated via the charge sharing. Once this analog computation is done across all the columns of BWPUs, the results are transferred to the MRU-Fs in which each contains SOGE-SA, shifter, and digital adder. SOGE-SA converts the analog values to digital values, and the shifter performs shifting considering their bit positions. The digital adder accumulates the shifted results and generates the final value in 20 bits.

During BP, the macro precharges R-RBL to V_{DD} and sets C-RBL to ground. Then, it injects the two input bits, which is an error in this case, through the backward wordline of MSB (BWLM) and LSB (BWLL). The post steps are similar to FP, except MWPUs on the same row are added across different columns through charge sharing of R-RBL to implement a transposed multiplication. In addition, the computation results are transferred horizontally to MRU-B.

6.3.2 CIMAT

Jiang et al. [2] propose a transpose SRAM-based computation-in-memory (CIM) architecture named CIMAT for multi-bit precision DNN training. The authors suggest three key architectural features to accelerate on-device training: 7T and 8T transpose SRAM bit-cell design, weight mapping strategies and data flow, and layer-level pipeline design for the training process. With them, CIMAT supports all four

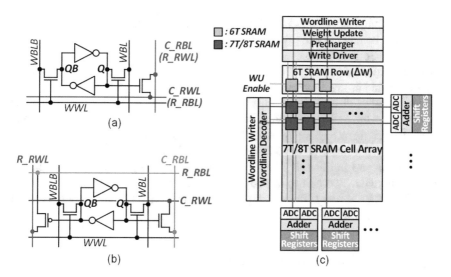

Fig. 6.3 (**a**) 7T transpose SRAM cell (**b**) 8T transpose SRAM cell (**c**) Overall architecture of CIMAT

stages of training (i.e., FP, BP, GC, and WU), unlike the TWT SRAM PIM only covers the first two. Modeled in a 7 nm CMOS technology, CIMAT successfully trains the ImageNet using the ResNet-18 model, achieving the energy efficiency of 10.79TOPS/W with the area of 121.51 mm^2. It is simulated based on NueroSim [4].

6.3.2.1 7T and 8T Transpose SRAM Cell Design

CIMAT proposes custom 7T and 8T transpose SRAM cell for in-memory processing while having standard 6T SRAM cells for generic usage as well. Figure 6.3a shows the proposed 7T SRAM cell, which allows bidirectional read, horizontal and vertical, and read-disturb-free access. It feeds the activation data to the weight data stored in the cell array differently according to its operation mode. During FP, column read wordline (C_RWL) is used for activation injection, and column-read-bitline (C_RBL) is used as a bitline for vertical partial sum read-out. The value of C_RBL becomes 1 only if both the weight bit on Q and the injected bit on C_RWL are 1. Therefore, the in-memory computation implements AND operation, and CIMAT uses this as a basic operation. For BP, their roles are changed. An error bit is injected through R_RWL, while R_RBL is used for horizontal partial sum read-out.

CIMAT also proposes an 8T SRAM cell that can execute FP and BP concurrently for higher throughput as shown in Fig. 6.3b. Occupying an additional area, the 8T SRAM cell design adds a PMOS transistor whose gate is connected to QB to support read accesses from both sides of the cell. In addition, R_RWL and R_RBL are added and connected to the PMOS transistor. C_RWL and R_RWL are used for activation and error input, respectively. Likewise, C_RBL and R_RBL are used for reading out partial sum by column and row, respectively.

Figure 6.3c shows the overall architecture of CIMAT, having a memory array based on the proposed 7T/8T SRAM cells with extra periphery circuits. To enable read/write and in-memory operation of the SRAM cell array, CIMAT has wordline writers in both directions for injecting activations via C_RWLs and errors via R_RWLs to the cells, a wordline decoder for writing weights to the cells, and a precharger for pre-charging write bitlines (WBLs) and bitline bars (WBLBs). To compute the result of the SRAM cell array, it has ADCs for the partial sum quantization and shifter and adder for accumulation of digital partial sums in bit-serial arithmetic. Two groups of periphery circuits exist at the bottom and the right side of the SRAM array. There is a special row of 6T SRAM cells at the top side of the SRAM array for WU.

6.3.2.2 Weight Mapping Strategies and Data Flow

CIMAT flattens a 3-d kernel in the direction of input channels and maps the flattened elements from different filters to different columns in a sub-array. Different locations of the elements in a filter are stored across different sub-arrays (e.g., 9 sub-arrays are required for $3 \times 3 \times N$ filters). CIMAT uses an adder tree to add the partial sums from different sub-arrays. To perform MAC operation using the in-memory operation, CIMAT pre-writes weights to the 7T/8T SRAM cells and injects activations through read-wordlines, C_RWL in Fig. 6.3. CIMAT adds the results of in-cell AND operations among different rows through the read bitlines and accumulates the results of different sub-arrays through the adder tree. Even though the adder tree can be an additional cost, it adopts this spatial accumulation scheme to make the FP and BP operation symmetric. Without this mapping, an accumulation result inside a sub-array is asymmetric because the result of FP is an entire partial sum while the result of BP is just part of the partial sum.

The transpose SRAM cell design of CIMAT removes the overhead of transposing the weight matrix of BP. With the proposed array design, the mapping scheme of the weight matrix to the SRAM array does not need to change. Instead, the accumulation direction should be reversed, from vertical to horizontal. With the same periphery circuits, the rest of the computation is the same as FP. With the proposed 8T SRAM design, FP and BP can be performed simultaneously within the same sub-array.

CIMAT uses extra non-transpose 6T SRAM arrays during GC to perform convolution operations between the error maps and corresponding activation maps. It first saves the error data in the SRAM array and loads the activation from the off-chip DRAM to the on-chip buffer. Each plane of error data is stretched into one column, and the next column stores the following output channel elements. CIMAT executes bit-wise multiplication and accumulation using the periphery circuits used in FP. The results of the columns form the entire gradient matrix. The multi-batch mode sends the gradients to off-chip DRAM to store the data. At the end of each batch, gradients are loaded back and accumulated on-chip.

After GC is done, each row of the gradient results is fetched to the additional 6T SRAM row residing above the 7T/8T SRAM. The data are fed into the shift registers row-by-row in a read-modify-write mode to multiply the learning rate. After that, the 6T row and a paired weight row of 7T or 8T are activated simultaneously. The subtraction of the two rows is done in the weight update module. To speed up this row-by-row data processing, CIMAT proposes an array-level pipelined architecture that updates different significant bits at different stages, as shown in Fig. 6.4.

6.3.2.3 Pipeline Design

In the case of 7T SRAM, CIMAT uses a 7-stage layer-level pipelining during FP and BP for achieving high throughput (Fig. 6.5a). Each stage computes multiple layers of a different image, while multiple images are processed throughout the pipelines

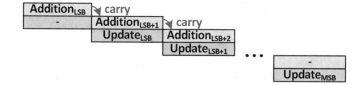

Fig. 6.4 Array-level pipeline design for weight update

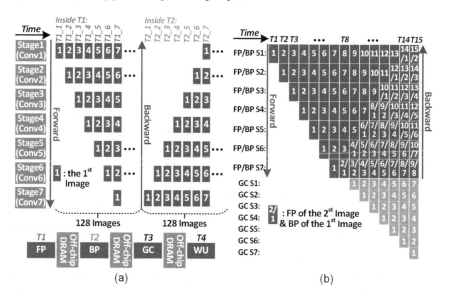

Fig. 6.5 Layer-level pipeline design (**a**) 7T SRAM (**b**) 8T SRAM

simultaneously. CIMAT balances out the execution times of stages by computing a different number of layers in each stage. The 7T SRAM design cannot execute FP and BP processes simultaneously because it only has a single read bitline. Either C_RBL or R_RBL acts as a read bitline, and the other acts as a read wordline, or vice versa. For example, when C_RBL is used for injecting input activation, C_RWL is used for partial sum read-out during FP. Therefore, FP and BP of one batch are computed serially, stage by stage.

On the other hand, 8T SRAM has two dedicated read bitlines, C_RBL and R_RBL. Using this, the BP processing of a previous image can be performed with the FP processing of a current image. For example, the box with 2/1 at time T8 means that the SRAM array is running both FP of the 2nd image and BP of the 1st image for the layer stage 7. This pipelined parallel execution of FP and BP is enabled by dual-bitline in the cell design and it boosts the throughput performance a lot. The generated intermediate data from both processes need to be saved off-chip for GC.

For a mini-batch that includes multiple images, GC generates gradients image by image. For example, if the mini-batch size is 128, 128 different weight gradients are generated after 128 runs. Finally, those gradients are averaged outside the chip, and the final WU is executed in one step inside CIMAT.

For the case of 8T SRAM, BP forms pipelines together with FP thanks to bidirectional readable cell design. As shown in Fig. 6.5b, the WG of the stage, which finishes both FP and BP, is also executed at duplicated CIM arrays with pipelining. By doing this, CIMAT saves energy consumption due to the reduction in off-chip memory access and on-chip standby leakage current.

6.3.3   HFP-CIM

Lee et al. [3] propose a heterogeneous floating-point computing-in-memory archi-tecture (HFP-CIM) by separately optimizing exponent processing and mantissa processing. The authors observe that most of the previous PIM works [1, 2] use fixed-point data due to the computational complexity of floating-point numbers. The complexity is mainly caused by the different operating characteristics between mantissa and exponent. For the multiplication of floating-point numbers, the exponent part can be done by simple addition and subtraction, while the mantissa part requires addition, subtraction, shifting, and leading-one-detection. Therefore, the computation cost of mantissa is much more expensive than that of the exponent, although the overheads of fetching and saving from/to memory are similar in both cases. Considering that PIM is specialized in computing simple operations in parallel due to its tight area and resource constraint, the PIM design that integrates both simple exponent computation and complex mantissa computation is worse than the one with only simple exponent computation from the performance perspective. This is why previous PIM works use simpler fixed-point as their data format.

However, using fixed-point numbers in ML training can harm its accuracy, which is a critical problem.

HFP-CIM suggests three key features to process floating-point numbers in PIM efficiently: (1) heterogeneous floating-point computing architecture and hardware design, (2) computing algorithm that reduces the communication between exponent and mantissa, and (3) data mapping and additional computing unit to support ML training. HFP-CIM is fabricated and verified in a 28 nm CMOS technology.

6.3.3.1 Heterogeneous Floating-Point Computing Architecture

As mentioned above, there may be a performance problem if we integrate both exponent and mantissa processing in the PIM hardware. Compute SRAM [5] supports both exponent and mantissa processing in a PIM macro. It has separate memory space for exponent and mantissa and performs processing for each by sharing the compute-in-memory logic in different time frames (i.e., time multiplexing), as illustrated in Fig. 6.6a. The proposed PIM macro suffers from a long latency in computing floating-point MAC. For instance, it requires more than 5000 cycles to compute 16-bit brain floating-point (BFP16) multiplications and a 32-bit floating-point accumulation using a single row of the macro. The proposed design tries to tackle a long latency issue by leveraging a high throughput. Each row performs the computation in parallel and gets multiple MAC results simultaneously. Nevertheless, the throughput degradation is still a problem since the utilization of macro rows is not always high. To address this problem, HFP-CIM decouples the exponent and mantissa processing and adopts PIM only for exponent operations. It proposes exponent computing-in-memory (ECIM) and mantissa processing engine (MPE). ECIM stores the exponents in memory and sends the shift amounts to MPE to reflect the difference of exponents. Then, each MPE performs shifting for both operands and multiplication and applies a normalization that configures the internal MAC result to a pre-defined floating-point format. The MPE returns a result of normalization to ECIM for exponent update in an exponent comparator. With this decoupled design that utilizes both in-memory and normal logic processing, HFP-

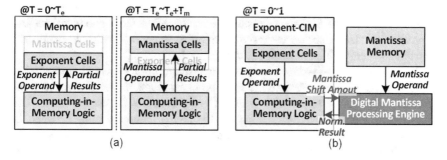

Fig. 6.6 (**a**) Conventional floating-point computing (**b**) Heterogeneous floating-point computing

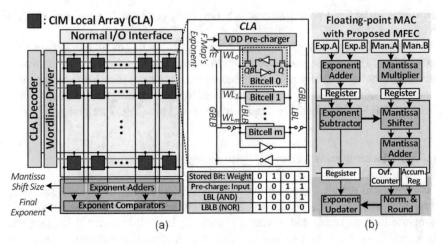

Fig. 6.7 (a) Overall architecture of ECIM (b) MFEC dataflow

CIM can execute floating-point MAC within two cycles. Figure 6.6b shows the concept of proposed heterogeneous floating-point computing.

Figure 6.7a shows the overall architecture of ECIM. It comprises CIM local arrays (CLAs), CLA decoder, wordline driver, normal I/O interface, and the peripherals for exponent computations. Its in-memory processing is composed of four steps: (1) storing the weight's exponent value in SRAM cells, (2) executing in-cell AND/NOR operation on local bitline (LBL) and bitline bar (LBLB) by enabling wordline of computing row after pre-charging LBL with exponent value of an input, (3) transferring the value of LBL and LBLB to global bitline bar (GBLB) and bitline (GBL) through drivers, and (4) loading the results in the global lines to peripheral circuits for further processing such as addition, subtraction, and comparison. During this process, the ECIM reduces power consumption with the following two features. First, it only precharges a single bitline, while the conventional memory precharges both bitlines (i.e., LBL and LBLB) to V_{DD} for read operation. ECIM either precharges LBL or LBLB depending on the input value. Second, it reuses the charge in GBL to reduce the switching power of ECIM. Not precharging every cycle, GBL reuses its charge from the previous cycle if the current cycle's in-cell AND/NOR result is the same. Since DNN computations tend to produce similar values on the results, exploiting temporal locality in memory is effective to energy-efficient hardware design.

Even though ECIM reduces the power consumption of exponent computation, mantissa computation in MPE still has two problems. First, the normalization result of mantissa has to be transferred to ECIM every cycle for exponent update. This process causes throughput degradation due to massive communication between the mantissa and exponent part. Second, the power consumption of mantissa's expensive arithmetic units such shifter and leading-one-detector is still high even after the ECIM is adopted. To solve these problems, HFP-CIM develops a mantissa-free-exponent-calculation (MFEC) algorithm, as shown in Fig. 6.7b. Unlike conventional floating-point MAC, MFEC does not normalize the mantissa result every cycle. Instead, it applies the normalization once after enough partial sum accumulation using a high-precision accumulator and an overflow detector, as there are many partial sums to be accumulated in DNN computation. By using MFEC, the processor minimizes the redundant normalization process of conventional floating-point MAC.

6.3.3.2 Overall Processor Design and Sparsity Handling

Figure 6.8a shows the overall processor design using HFP-CIM, composed of the top RISC controller, multiple instances of heterogeneous-exponent-mantissa-training-core (HEMTC), aggregation & activation core (AAC) that accumulates

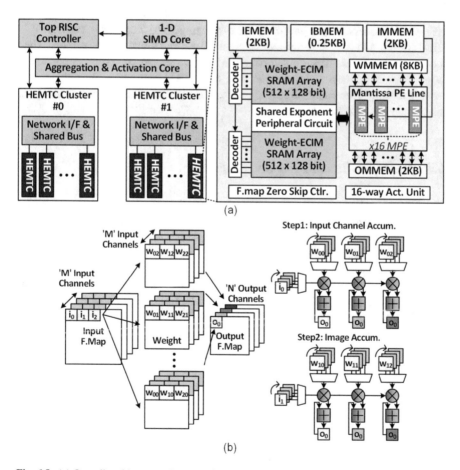

Fig. 6.8 (a) Overall architecture of proposed processor (b) Data mapping

partial sums from HEMTCs, and 1-D SIMD core computing element-wise multi-plication. The HEMTC contains the ECIM macro and MPEs. Utilizing these units, the processor supports the ML training composed of FP, BP, GC, and WU. It devises a data mapping that can additionally support the convolutional layer as well as maximally reuse partial sums for MFEC.

Figure 6.8b shows the data flow and data mapping of the proposed processor. There are two kinds of accumulation, and the first one is input channel accumulation. Each column of CLAs in ECIM contains the weight of different output channels, which is first flattened in the input channel direction. These values between all the rows are accumulated. After the input channel accumulation, the weight value in ECIM is changed to another element of the kernel window, the corresponding input value is injected, and the same accumulation process is performed. This process is image accumulation, and during these two steps, the partial sums are accumulated in the same memory space. This paper does not describe how all the processor executes numerous functions of ML training and movement of intermediate data.

As previous works [6, 7] exploit the sparsity of data for high energy efficiency, HFP-CIM processor also performs zero-skipping. The zero skip controller in HEMTC gets non-zero encoded exponent, mantissa, and bitmap from memory. Then it converts them into non-zero value feature maps with their corresponding indexes, saves them in queues, and feeds the data into ECIM and MPE. With these data feeding schemes, the HEMTC reduces energy consumption and latency by skipping the calculation of zero values.

6.4   ReRAM-Based PIM for Training

Metal-oxide resistive random access memory (ReRAM) stores bit information by changing the resistance of the cell. ReRAM has recently received much attention as a suitable non-volatile memory for PIM, because it can store a large DNN model with fast read speed and perform efficient matrix-vector multiplication in a crossbar structure. Recent PIM works proposed ReRAM-based CNN inference accelerators to overcome the memory bandwidth bottleneck [8, 9]. Some other works have aggressively adopted ReRAM technology for ML training to show its possibility [10, 11]. This section describes ReRAM-based PIM accelerators that support ML training as well as inference.

6.4.1   PipeLayer

Song et al. [10] propose PipeLayer, a ReRAM-based PIM accelerator for CNN. The authors claim that previous ReRAM-based PIM works, PRIME [8] and ISAAC [9], cannot support training due to a few reasons. Both works do not consider the complex data dependency of training. Unclear data organization and data mapping of PRIME make it hard to handle different data movements in FP and BP of training. ISAAC's deep pipelining for increasing system throughput is only effective when there are sizable consecutive inputs, but that is not the case in the training. Moreover, its pipeline design is vulnerable to bubbles and stalls. PipeLayer suggests a simple intra-layer and inter-layer pipeline design that executes both inference and training operations to overcome the above problems. In addition, unlike other conventional ReRAMs, PipeLayer adopts a spike-based input/output signaling scheme rather than a voltage-level-based. This scheme eliminates the overhead of DACs and ADCs. PipeLayer is simulated based on NVSim [12] and its energy and area model use the measurement results from Niu et al. [13] and Fackenthal et al. [14].

6.4.1.1 Architecture of PipeLayer

PipeLayer exploits ReRAM cells to perform computations without using other processing units. The proposed design is divided into morphable sub-arrays (Morps) and memory sub-arrays (Mems). Morp performs computation and stores the data, while Mem is only used for data storage. For the DNN training, each Morp computes a layer while Mem stores both the activation data that propagates to the next layer and the intermediate data generated during FP that are necessary for BP, GC, and WU. The role of Morp is interchangeable; PipeLayer uses Morp as a computation unit during inference for high throughput or as a Mem during training for storing intermediate data. PipeLayer handles mini-batching by averaging the partial derivatives if the batch size B is larger than one. Mem accumulates the partial derivatives of a mini-batch and sends the values to Morp, and Morp reduces the magnitude of input spikes by B.

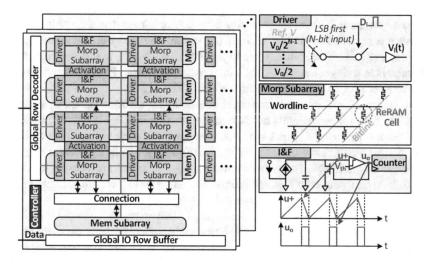

Fig. 6.9 Architecture overview of pipeLayer

Figure 6.9 shows the overall architecture of PipeLayer. Note that it does not include any processing units for computations as the ReRAM cell arrays substitute them. The in-memory computation of PipeLayer is similar to that of ISAAC. It uses a weight spike coding scheme for input/output signaling to remove the overhead of ADCs and DACs. Unlike ISAAC, which still needs ADCs for output spikes, PipeLayer does not need both DACs and ADCs thanks to the spike driver and integration-and-fire circuits, respectively. When the input is N bit, the spike driver iterates N times and generates a sequence of weighted spikes by looking up reference voltage at each cycle. Then, it feeds them to the ReRAM cell array (i.e., Morp). The weight data is stored in the ReRAM cell array as cell conductance, and cells are located at the cross points of the wordlines and bitlines. Since the multiplication result of conductance and voltage is a current value, a current flowing at each cell can be viewed as a multiplication result of the input value from the spike driver and the weight value in the cell. Then, PipeLayer accumulates the in-cell multiplication results by sharing the current on a bitline. After accumulation, it uses integration-and-fire (I&F) unit, which integrates input current and generates output spikes. Furthermore, the counter connected to the output spikes finally converts the spikes to digital values. For the network that needs high resolution, it accumulates the partial sums after shifting.

6.4.1.2 Data Mapping and Parallelism of PipeLayer

As shown in Fig. 6.10a, PipeLayer flattens weight kernels and stores them in Morp. Each column of the cell array includes the weights from a flattened kernel. PipeLayer feeds the input data after flattening them with the same method as the kernel's. For the input case whose width and height are set to 114, PipeLayer feeds the input data through wordlines and accumulates the multiplication results through bitlines. Because it takes only a single cycle for element-wise multiplications between a flattened input and each of the weights by using all the cross points, it needs $112 \times 112 (=12544)$ cycles to finish all the outputs. Since the mapping of all kernels to a single huge ReRAM sub-array is unrealistic, PipeLayer partitions it into smaller ReRAM sub-arrays with a size of 128×128.

To improve performance, PipeLayer can compute multiple flattened inputs at the same time for the weights in the same layer. This strategy is called intra-layer parallelism. To this end, PipeLayer needs to store the same weight data G times (Fig. 6.10b). For an extreme case, the results of the layer could be generated in just one cycle if G is 12544. The authors chose the G value to 256, considering the linear increase in hardware cost. Since it is only simulation-based, we are not certain that PipeLayer can be efficiently implemented.

In addition, PipeLayer exploits inter-layer parallelism where it computes multiple layers from different images in a mini-batch in parallel. For this parallelism, the proposed design does a data computation pipeline from different images that do not have data dependency between them. Figure 6.10c compares a conventional design and the proposed strategy. In the conventional design, the computation is

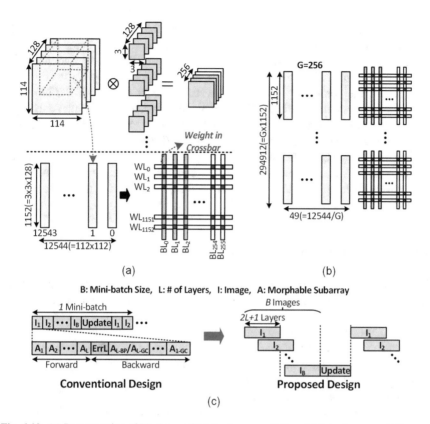

Fig. 6.10 (**a**) Data mapping of PipeLayer (**b**) Intra-layer parallelism (**c**) Inter-layer parallelism

sequential and has a long latency since there is no pipelining. The pipelining design of PipeLayer increases the throughput by computing different images at different Morp concurrently. However, the PipeLayer needs more buffers to enable this high-performance pipelining design, which causes extra area overhead.

6.4.2 FloatPIM

Imani et al. [11] propose FloatPIM, a ReRAM-based in-memory accelerator for DNN training with floating-point data type. The authors pointed out three critical problems for the previous ReRAM-based PIMs that support DNN training. First, they are bounded to fixed-point precision because a floating-point number requires more multi-bit memristors to represent a value. This constraint causes an accuracy drop during training. Second, the usage of ADC/DAC blocks has a considerable overhead in chip area and power. Third, the multi-bit memristor is not sufficiently reliable for commercialization, unlike single-level NVMs [15].

FloatPIM supports both floating-point and fixed-point computation by using the basic NOR operations of single-bit devices. Unlike other ReRAM-based PIMs using analog computation, FloatPIM uses digital computation. Thus, it does not need area and power-hungry ADC/DAC blocks. The top-level operation process comprises two phases: the computing phase and the data transfer phase. All the blocks of FloatPIM compute the matrix multiplication and convolution in parallel during the computing phase. Each block transfers the result row to its neighbor block in a pipelined manner in the data transfer phase. FloatPIM is synthesized using System Verilog and Design Compiler. It is evaluated with a custom cycle-accurate simulator, circuit-level simulators including HSPICE, and mathematical models. Although FloatPIM is the latest ReRAM-based PIM architecture supporting the whole process of DNN training, its feasibility to real hardware is not clear.

6.4.2.1 FloatPIM's Digital Operation

Unlike conventional memristor processing leverages fast and energy-efficient analog computation with ADC/DAC blocks, FloatPIM computes only in the digital domain and stores the values directly to the cells. Hence, it does not need any ADC/DAC blocks and sense amplifiers that incur hardware overhead. Figure 6.11a shows the digital computation in FloatPIM. The state of an output device switches between the two resistive states, R_{ON} (low resistance or logical 1) and R_{OFF} (high resistance or logical 0), whenever the voltage across p and n terminals exceeds a threshold. FloatPIM uses this property to implement the NOR gate in a memory.

In the beginning, only the output device is initialized to R_{ON} while the other inputs are set to R_{OFF}. Then the execution voltage V_0 is applied at the p terminals

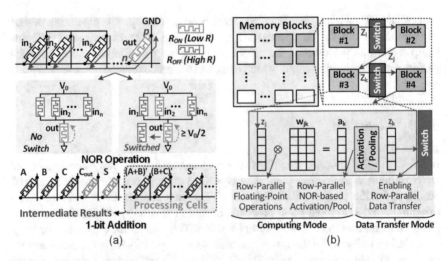

Fig. 6.11 (a) Digital PIM operations (b) Overview of FloatPIM

of the inputs. If the state of all parallel-connected input devices is R_{OFF}, the state of the output device does not change. However, if one of the input devices' state changes to R_{ON}, the output memristor is switched from R_{ON} to R_{OFF}. Since an assertion among the input devices causes the output de-asserted, it implements the NOR operation. As NOR operation is functionally complete, other arithmetic operations such as addition and multiplication can be implemented. Each row contains cells for storing both operands to read simultaneously and separate processing cells only for storing intermediate results. In-memory operation of ReRAM-based PIM is slower than CMOS-based PIM due to the slow switching speed of the memristor. To overcome this, FloatPIM suggests even more parallelism during computation. It can compute addition and multiplication in parallel, irrespective of the number of rows.

Figure 6.11b depicts the overview of FloatPIM. It is composed of crossbar memory blocks, in which each block contains data from a different layer of DNN. The memory blocks store only weight data during inference. However, during training, they store weights, activation gradients (derivatives of activation functions), and results of activation functions. Each block sends computation results to the next block through the switch that aligns the data structure for the data transfer phase.

6.4.2.2 Hardware Architecture

Figure 6.12 shows the hardware architecture of FloatPIM. It comprises 32 tiles, in which each tile contains 256 memory blocks, and each memory block contains 1k × 32k data. FloatPIM reads and writes data in a row-parallel way. To enable this, it utilizes the switches during data communication between blocks. When FloatPIM

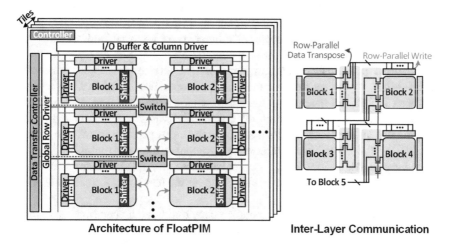

Fig. 6.12 Hardware architecture of FloatPIM

reads a vector from the block, the switch connects each row data point to each driver's column for the write operation to the next block. The shifter is inside the memory block to support convolution operation. The controller computes the loss function and controls data drivers and switches.

FloatPIM uses different parallelism schemes in FP and BP. For FP, it computes each batch in each tile at the same time. For BP, the FloatPIM has 2 configurations: low-power FloatPIM (FloatPIM-LP) and high-power FloatPIM (FloatPIM-HP). It determines the parallelism strategy considering the trade-off among speed, energy efficiency, and memory size. In the FloatPIM-LP, a single memory block iteratively computes all data points in a mini-batch, generates gradients, and subtracts generated gradients from the current weights. In contrast, in the FloatPIM-HP, multiple blocks compute different data points in a mini-batch in parallel, sum up the gradients across different blocks, and update the weights. Even though FloatPIM-HP performs computation faster, the weights need to be duplicated in each block. This duplication makes FloatPIM-HP consume large memory and energy.

6.4.2.3 Training of FloatPIM

During FP, FloatPIM processes the input data in a pipeline stage. While the value of a single batch passes through each data point, FloatPIM stores gradients and the results of activation functions in each data point. For BP, FloatPIM measures the loss function in the last output layer and updates the weights of each layer using the previously stored data, while error propagates each data point.

Figure 6.13a shows how FloatPIM performs two key operations of CNN: matrix-vector multiplication and convolution operation. FloatPIM stores multiple copies of the input vector in multiple rows and stores the weight matrix in the transposed shape for matrix-vector multiplication. It first performs the multiplication between the inputs and weights and then accumulates the multiplication results horizontally. While FloatPIM performs the computation in a row-parallel way for high performance, it always needs extra memory for input vector copy, which is memory area overhead.

FloatPIM performs convolution using weight interconnect logic, which is a barrel shifter. It prevents frequent memory write operation, which is a considerable overhead in ReRAM-based PIM since its memory write operation is slow. During the convolution, FloatPIM stores all convolution weights in a single row and copies them to other rows. It first multiplies the corresponding inputs and weights considering the convolution window and computes the next multiplication with shifted inputs. Then, it performs accumulation with the results for the final result. Specifically, for the $N \times N$ convolution window, the number of the shift operations is $N - 1$.

Since FloatPIM performs frequent data copy during both operations, it supports an optimized data copy operation, which writes the same value to all rows within two cycles. FloatPIM supports the Sigmoid function by using three terms of the Taylor expansion and the ReLu function for the activation. The max/min pooling

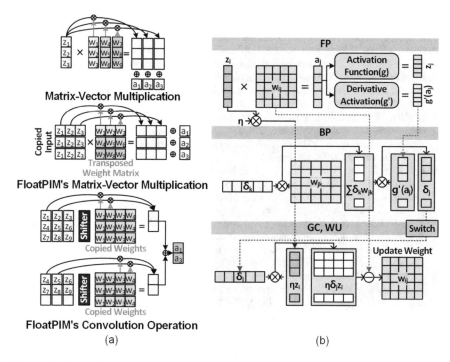

Fig. 6.13 (**a**) Matrix-vector multiplication and convolution (**b**) Training of FloatPIM

of FloatPIM first compares the exponent and then compares the mantissa with the same maximum/minimum exponent.

During BP, GC, and WU, the error vector propagates to corresponding memory blocks to access required data points for weight update. For the BP, FloatPIM multiplies the copied error vector with the transposed weight matrix and multiplies the activation gradient stored during FP. The resulting error propagates to the next memory block. To perform GC, FloatPIM multiplies the same copied error vector to the result of the activation function scaled with the learning rate and finally updates the weight matrix. The whole process of CNN training in FloatPIM is summarized in Fig. 6.13b.

References

1. J.-W. Su, X. Si, Y.-C. Chou, T.-W. Chang, W.-H. Huang, Y.-N. Tu, R. Liu, P.-J. Lu, T.-W. Liu, J.-H. Wang, Z. Zhang, H. Jiang, S. Huang, C.-C. Lo, R.-S. Liu, C.-C. Hsieh, K.-T. Tang, S.-S. Sheu, S.-H. Li, H.-Y. Lee, S.-C. Chang, S. Yu, and M.-F. Chang, 15.2 a 28 nm 64Kb inference-training two-way transpose multibit 6T SRAM Compute-in-Memory macro for AI edge chips, in *2020 IEEE International Solid-State Circuits Conference-(ISSCC)*. IEEE, Piscataway (2020), pp. 240–242

2. H. Jiang, X. Peng, S. Huang, S. Yu, CIMAT: a compute-in-memory architecture for on-chip training based on transpose SRAM arrays. IEEE Trans. Comput. **69**(7), 944–954 (2020)
3. J. Lee, J. Kim, W. Jo, S. Kim, S. Kim, H.J. Yoo, *ECIM: Exponent Computing in Memory for an Energy Efficient Heterogeneous Floating-Point DNN Training Processor*. IEEE Micro (2021)
4. P.Y. Chen, X. Peng, S. Yu, NeuroSim: a circuit-level macro model for benchmarking neuro-inspired architectures in online learning. IEEE Trans. Comput.-Aided Design Integr. Circuits Syst. **37**(12), 3067–3080 (2018)
5. J. Wang, X. Wang, C. Eckert, A. Subramaniyan, R. Das, D. Blaauw, D. Sylvester, 14.2 a compute SRAM with bit-serial integer/floating-point operations for programmable in-memory vector acceleration, in *2019 IEEE International Solid-State Circuits Conference-(ISSCC)*. IEEE, Piscataway (2019), pp. 224–226
6. J.H. Kim, J. Lee, J. Lee, H.J. Yoo, J.Y. Kim, Z-PIM: An energy-efficient sparsity aware processing-in-memory architecture with fully-variable weight precision, in *2020 IEEE Symposium on VLSI Circuits*. IEEE, Piscataway (2020), pp. 1–2
7. J. Yue, Z. Yuan, X. Feng, Y. He, Z. Zhang, X. Si, R. Liu, M.-F. Chang, X. Li, H. Yang, Y. Liu, 14.3 A 65 nm computing-in-memory-based CNN processor with 2.9-to-35.8 TOPS/W system energy efficiency using dynamic-sparsity performance-scaling architecture and energy-efficient inter/intra-macro data reuse, in *2020 IEEE International Solid-State Circuits Conference-(ISSCC)*. IEEE, Piscataway (2020), pp. 234–236
8. P. Chi, S. Li, C. Xu, T. Zhang, J. Zhao, Y. Liu, Y. Wang, Y. Xie, Prime: a novel processing-in-memory architecture for neural network computation in ReRam-based main memory. ACM SIGARCH Comput. Archit. News **44**(3), 27–39 (2016)
9. A. Shafiee, A. Nag, N. Muralimanohar, R. Balasubramonian, J.P. Strachan, M. Hu, R.S. Williams, V. Srikumar, ISAAC: a convolutional neural network accelerator with in-situ analog arithmetic in crossbars. ACM SIGARCH Comput. Archit. News **44**(3), 14–26 (2016)
10. L. Song, X. Qian, H. Li, Y. Chen, Pipelayer: a pipelined reram-based accelerator for deep learning, in *2017 IEEE International Symposium on High Performance Computer Architecture (HPCA)* (pp. 541–552). IEEE, Piscataway (2017)
11. M. Imani, S. Gupta, Y. Kim, T. Rosing, FloatPIM: in-memory acceleration of deep neural network training with high precision, in *2019 ACM/IEEE 46th Annual International Symposium on Computer Architecture (ISCA)*. IEEE, Piscataway (2019), pp. 802–815
12. X. Dong, C. Xu, Y. Xie, N.P. Jouppi, NVSim: a circuit-level performance, energy, and area model for emerging nonvolatile memory. IEEE Trans. Comput.-Aided Design Integr. Circuits Syst. **31**(7), 994–1007 (2012)
13. D. Niu, C. Xu, N, Muralimanohar, N.P. Jouppi, Y. Xie, Design trade-offs for high density cross-point resistive memory, in *Proceedings of the 2012 ACM/IEEE International Symposium on Low Power Electronics and Design* (2012), pp. 209–214
14. R. Fackenthal, M. Kitagawa, W. Otsuka, K. Prall, D. Mills, K. Tsutsui, J. Javanifard, K. Tedrow, T. Tsushima, Y. Shibahara, G. Hush, 19.7 A 16 Gb ReRAM with 200 MB/s write and 1 GB/s read in 27 nm technology, in *2014 IEEE International Solid-State Circuits Conference Digest of Technical Papers (ISSCC)*. IEEE, Piscataway (2014), pp. 338–339
15. Intel, Intel and Micron Produce Breakthrough Memory Technology (2015). Available via DIALOG. https://newsroom.intel.com/news-releases/intel-and-micron-produce-breakthrough-memory-technology/

Chapter 7
PIM Software Stack

Donghyuk Kim and Joo-Young Kim

This chapter will discuss a software stack for PIM and the challenges that must be overcome to be well-suited in the conventional computer architecture. In a standard term, a software stack consists of layers of software components that create a complete platform without any additional component to support applications. It provides an interface between hardware and programmers through the layers of application, framework, library, runtime, and device driver. In order to efficiently adopt PIM into the conventional architecture, a software stack needs modification on these layers, as shown in Fig. 7.1.

The primary purpose of the PIM software stack is not just to make an application run on the PIM hardware; it must be optimized for the PIM hardware in a seamless manner, regarding the utilization of the hardware, scheduling, and optimization of the code. Also, it must aim for high programmability and optimization to a variety of applications and architecture systems, which should make PIM hardware convenient to programmers and system architects.

7.1 PIM Software Stack Overview

In order to adopt PIM properly, we must thoroughly tackle the entire software architecture layers, including PIM application, PIM library, and PIM device driver. An application is a high-level software code that users write. However, not every type of application could exploit PIM well. Due to the high internal bandwidth

The original version of the chapter has been revised. A correction to this chapter can be found at https://doi.org/10.1007/978-3-030-98781-7_9.

D. Kim · J.-Y. Kim (✉)

School of Electrical Engineering (E3-2), KAIST, Daejeon, South Korea

e-mail: kar02040@kaist.ac.kr; jooyoung1203@kaist.ac.kr

J.-Y. Kim et al. (eds.), *Processing-in-Memory for AI*, https://doi.org/10.1007/978-3-030-98781-7_7

143

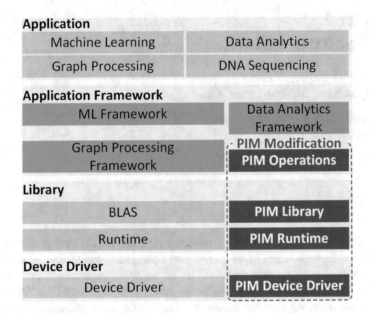

Fig. 7.1 Software stack with PIM modification support

of PIM with limited computation capability, only certain types of applications could run better on PIM than other computation platforms such as CPU, GPU, and accelerators. For example, machine learning is a type of application that can benefit from PIM if appropriately applied. Especially, memory-intensive applications such as fully connected layers can take advantage of PIM by supporting enough internal memory bandwidth. One thing to notice is that an application itself may be hardware-agnostic (i.e., does not know the PIM hardware property), missing the chance of fully exploiting PIM resources. It needs a PIM library to do that job; PIM library makes a seamless transition from a high-level application to the actual execution of PIM. It has three essential roles. First, it identifies codes that could run on PIM. Second, it generates PIM instructions identified as to be run on PIM and prepares operand data for the PIM instructions. Third, it executes the PIM kernel with the PIM instructions. After obtaining appropriate instructions and data with the PIM library, the PIM instruction requests are sent to the PIM device driver. Then the PIM device driver reserves memory space for data and instructions and offloads them to a memory controller.

A recent research paper of HBM-PIM [1] gives an example of a complete PIM software stack. Figure 7.2 depicts the overview of the modified software stack of HBM-PIM for PIM adoption. Before diving into its software stack, we need a brief explanation of the microarchitecture of the HBM-PIM execution unit. As shown in Fig. 7.3, there are a scalar register file (SRF), general register files (GRF_A and GRF_B), a command register file (CRF), and a 16-wide single instruction multiple data (SIMD) floating-point unit (FPU). The SIMD FPU consists of a pair of 16 FP16 multipliers and adders. It is controlled by the RISC-style 32-bit instructions stored in

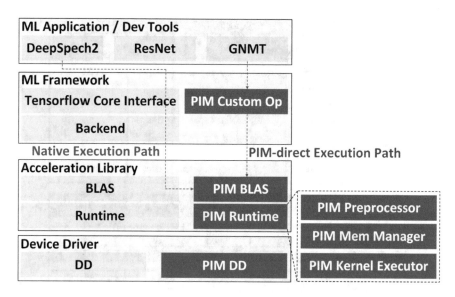

Fig. 7.2 HBM-PIM software stack

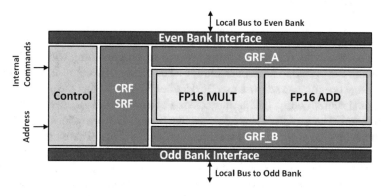

Fig. 7.3 HBM-PIM microarchitecture

CRF, while the host is responsible for providing instructions to each PIM execution unit. HBM-PIM's instructions and instruction format are illustrated in Fig. 7.4.

The software stack shown in Fig. 7.2 is modified to utilize the PIM execution unit efficiently. It supports basic linear algebra subprograms (BLAS), runtime, and a device driver to allow users to run the original source code without any modifications. As another option of programming, it also supports PIM custom operations that directly invoke the PIM hardware. They are PIM BLAS based TensorFlow operations: addition (ADD), multiplication (MUL), rectified linear unit (ReLU), long short-term memory (LSTM), general matrix-vector multiplication (GEMV), and batch normalization (BN). They explicitly call the corresponding PIM BLAS library. Then, the PIM BLAS calls the PIM kernel to generate PIM micro-kernel

PIM Instructions				
Op. Type	Operand (SRC0)	Operand (SRC1)	Result (DST)	# of Combinations
MUL	GRF, BANK	GRF, BANK, SRF_M	GRF	32
ADD	GRF, BANK, SRF_A	GRF, BANK, SRF_A	GRF	40
MAC	GRF, BANK	GRF, BANK, SRF_M	GRF_B	14
MAD	GRF, BANK	GRF, BANK, SRF_M SRF_A (for SRC2)	GRF	28
MOV (ReLU)	GRF, BANK	-	GRF	24

PIM Instruction Format															
31	30	29	28	27	26	25	24	23	22	21	20	19	18	17	16
OPCODE					U									IMM0	
OPCODE				DST			SRC0					U			
OPCODE				DST			SRC0			SRC1			SRC2		
15	14	13	12	11	10	9	8	7	6	5	4	3	2	1	0
IMM0										IMM1					
U		R	U		DST #		U		SRC0 #		U		SRC1 #		
A	U		U		DST #		U		SRC0 #		U		SRC1 #		

Fig. 7.4 HBM-PIM instructions and format

codes and execute them. This path allows users for manual and direct use of the PIM execution unit. In the case of not using the manual programming with PIM custom operations, PIM runtime is responsible for a seamless connection. It optimizes the TensorFlow operations and invokes a PIM kernel without having the user modifying the original source code. PIM runtime consists of a pre-processor, memory manager, and executor. The pre-processor analyzes the TensorFlow operations to find which operations to offload to the PIM execution unit at runtime. The memory manager manages the PIM operations and also maps PIM micro-kernel code and operand data to the memory space allocated by the PIM device driver. The important thing is to match the data location and the execution unit to minimize the data movement overhead. The PIM executor calls and configures the PIM kernel. PIM device driver reserves memory for the PIM execution unit. It forces the reserved memory space to be uncacheable to guarantee the DRAM memory accesses from the host processor to PIM. It also manages cache coherence issues between the host and the PIM execution unit by not using caches for the shared data.

The programming model for the PIM execution unit is to execute the PIM micro-kernel with a valid memory request to DRAM. The programming model is depicted in Fig. 7.5. The most important aspects of the PIM kernel are utilizing the whole internal HBM-PIM compute bandwidth and ensuring the order of the memory requests to keep the execution order of the PIM micro-kernel. First, to fully utilize the compute bandwidth of the HBM-PIM, it generates enough threads to map kernels to all of the PIM execution units. Each thread can send a memory request, and there should be enough number of threads to utilize the GRF accessing size, which is 256B in total. Some threads are grouped and form a thread group

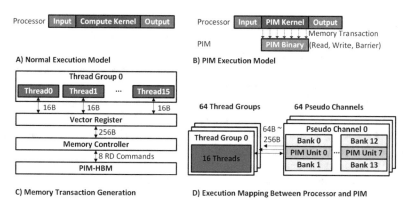

Fig. 7.5 HBM-PIM programming model

that is mapped to an HBM pseudo-channel (pCHs). For example, HBM-PIM uses a processor with an ISA memory access size of 16B. Then, to match the GRF accessing size of 256B, 16 threads, each of which has a memory access size of 16B, are grouped in a thread group. There should be 64 thread groups for HBM-PIM with 64 pCHs. Second, it must ensure the order of the memory request since it correlates to the order of the execution of HBM-PIM and its correctness. To ensure the order of memory requests in the same thread group, HBM-PIM uses barrier application programming interface (API). The barrier API forces the PIM kernel to preserve the order of the requests. HBM-PIM also resolves fence overhead between different DRAM channels by one-to-one mapping of each thread group to each DRAM channel. In between each thread group, it is programmed that each thread group can only access a single DRAM channel, which reduces the requests overhead between each thread group.

7.1.1 PIM Software Stack Challenges

The layers of the software stack must be modified to program PIM efficiently. However, modifying the software stack for PIM introduces several challenges. First, we must identify PIM offloading execution. There are types of applications or executions that could run better on PIM. It is important to know what properties they have and how to distinguish them from the rest. Second, data mapping must be appropriately managed to reduce the data movement overhead. Mapping data in an appropriate format is crucial for some types of PIM hardware. If data mapping is properly done, it could relieve the internal data movement bottleneck. Third, scheduling of the PIM executions must be applied for efficiency and the best performance of PIM. It helps to optimize the utilization of the PIM and the host processor by concurrently offloading execution kernels to both at the same time. To schedule executions to appropriate hardware, we need to look at the information of the list of executions and runtime resource utilization of each part of the hardware. Fourth, a cache coherence issue must be resolved for shared data on PIM and the host processor. Cache coherence issue occurs when multiple cores with their own local cache try to access the same address data in memory. Each core might not look at the same data value because the data might be modified in one core's cache but not in the others. This issue occurs the same for PIM and the host processor. The host processor could still look at the stale data if the PIM changes a data value in its local cache. There are conventional cache coherence protocols to solve this issue. However, these poorly work for PIM due to the narrow off-chip bandwidth between PIM and the host. The above challenges must be addressed with appropriate solutions. The following sections discuss the challenges and recent research that propose possible solutions.

7.2   PIM Offloading Execution

The first challenge is to identify what codes to be executed on PIM. Since not every code is efficient running on PIM, we need to distinguish a PIM-friendly code that effectively exploits PIM. PIM-friendly codes can be statically assigned to PIM cores by programmers manually. It should come with a deep understanding of the PIM architecture, code property, and the benefit of offloading codes to PIM. Depending on what and how the PIM execution unit is designed in memory, deciding what code to offload to PIM varies. Identifying PIM-friendly codes is relatively straightforward for PIM cores with a custom logic if the PIM logic is specialized for a certain function. For example, a PIM core in Newton [2] consists of 16 multipliers and a reduction adder tree with a fixed data flow. This PIM architecture is specifically designed for a particular application, a matrix-vector multiplication in this case. Utilizing Newton for matrix-vector multiplications results in promising performance with its specialized logic unit in each DRAM bank. It gains wide internal bandwidth to each PIM core while reducing the amount of data transferred from the memory to the processor, which eases the burden on memory bottleneck issues. The energy consumption on off-chip data transfer is significantly reduced, and the throughput is increased by utilizing higher internal DRAM bandwidth.

On the other hand, identifying PIM-friendly codes in general-purpose PIM cores is much more difficult. It must be analyzed that the code is memory-intensive, which means there is a memory bottleneck on off-chip bandwidth between the memory and the processor. Typically, PIM copes with memory-intensive but low data locality applications. Conversely, CPU has an advantage in compute-intensive and cache-friendly applications. Such memory-intensive applications require tremendous data from memory. If the processor can only use the same data for few times and needs to request a lot of data from memory more than a given off-chip bandwidth, memory bottleneck happens. Boroumand et al. [3] propose an efficient tool flow rather than leaving identifying memory-intensive PIM-friendly code to programmers' manual work. They analyze certain conditions of target applications by hardware performance counters and the energy model. They suggest four criteria that the code is PIM-friendly if (1) it consumes the most energy out of all functions in the workload, (2) its data movement consumes a significant fraction of the total workload energy, (3) the last cache misses per kilo instruction (MPKI) is greater than 10, and (4) data movement is the single most significant component of the function's energy consumption.

This research demonstrates applying its tool flow in analyzing primary Google consumer workloads, including the Chrome browser, TensorFlow mobile, video playback, and video capture. For example, it evaluates the TensorFlow mobile machine learning (ML) inference application with two candidates: packing and quantization. Packing is a type of function to pre-process the matrix for GEMM operations. It re-organizes the order of tiles in matrices to maximize the cache locality, which causes considerable data movement. Quantization is a function to convert 32-bit data type into 8-bit data type to minimize the computation. It is applied twice to the input and output of every 2-d convolution. It scans minimum and maximum values of data and does simple multiplication and addition for each data. The primary purpose of these two functions in ML is to help minimize the energy consumption and execution time during inference, but they do not seem to serve their purpose well on the CPU. With frequent occurrences of the functions in executing a whole application, the functions themselves cause significant overhead in external memory access. However, the two functions satisfy the criteria of PIM-friendly code by the fact that they are memory-intensive operations with low computation overhead.

Figure 7.6 shows the breakdown of energy consumption and execution time for various ML models on CPU. The two candidate functions for PIM, packing and quantization, account for 39.3% of total system energy and 54.4% of data movement energy between the CPUs and main memory. For the execution time analysis, they appear to spend 27.4% of total execution time. Additionally, both candidates do not require complicated computations so that PIM cores can handle them with minor adjustments. Figure 7.7 shows the difference between the quantization flow on CPU-only and CPU+PIM. Utilizing PIM minimizes the data movement overhead by computing all the simple but memory-intensive quantization processes in memory. This research shows the PIM cores running both functions reduce the power consumption by 50.9% and the execution time by 57.2%. Most of the reduction in power consumption comes from reduced data movement. Also, exploiting PIM cores in doing such functions provides greater internal memory bandwidth, low data access latency, and also enables parallel execution of GEMM in CPU while PIM can take care of the offloaded functions.

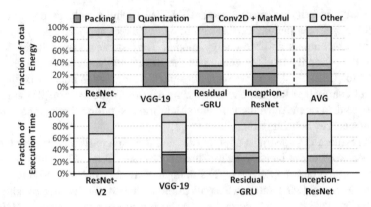

Fig. 7.6 Energy and execution time breakdown on various models on CPU

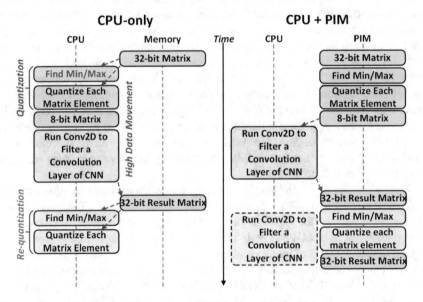

Fig. 7.7 Comparison on quantization flow of CPU-only and CPU+PIM

7.3 PIM Data Mapping

The second challenge is to manage data mapping for efficient programming. Data mapping strategy must consider the PIM hardware architecture and its target application. Inappropriate data mapping scheme to the memory generates an even worse data movement bottleneck in the system and degrades the overall performance with inefficient data access patterns by PIM computation units. The best data mapping strategy must be optimized differently for different types of PIM architectures and applications since it affects the computing performance of PIM.

Data mapping strategy depends on the size of the data granularity of PIM computation units. Data granularity varies along with the location of PIM computation units across different architecture levels, from DRAM's subarray level to bank level. They could be placed inside each DRAM's bank and access only selected data after a column decoder or before a column decoder with a whole row of the subarray. For example, one type of PIM hardware is located in DRAM's subarray level, and PIM units have data granularity of an entire row [4–6] with bulk-bitwise operations. The issue here is that they require data to be aligned in the same row and located in a certain subarray in order for PIM to execute accurately. It can be managed by generating sequentially aligned physical addresses from given virtual addresses in the operating system and exposing subarray's address information to the memory controller. These approaches ensure that the data can be physically located in a specific DRAM subarray within the same row.

Another type of PIM hardware is a bank-level PIM, where each PIM core is located in each bank, and computation is done after the column decoder. Bank-level PIM has fewer issues in aligning data since it requires a smaller size of data granularity. However, the biggest issue comes from the irregular data access pattern. Unlike CPU, bank-level PIM is limited to have caches or registers where some amount of data can be held locally. Consequently, data access time is dominated by accessing memory cells rather than each PIM core's local registers. Additionally, the relative distance between the bank that stores the target data and the PIM computation unit that can be either in the same bank or another bank causes overhead in data movement. First, a sequential memory accesses pattern guarantees the shortest data read latency within a bank in DRAM. While it requires additional row-to-row delay with DRAM's pre-charge and activation commands in accessing different row address data, a sequential memory access pattern can read the same row without the additional delay. Second, memory access from a PIM core to its neighbor bank memory burdens the global data bus. Many memory requests from different PIM cores potentially cause bottlenecks between banks. This inter-bank data movement can be done through a global data bus, if the PIM architecture supports its bank-to-bank data transmission. Otherwise, it must be done by the memory copy function, which moves data all the way from the source memory address to the host and back to the destination memory address. A solution to this issue can come from optimizing data mapping and assigning the proper PIM core to execute with the corresponding data. By matching data location and the execution of code to a specific PIM core, it is possible to alleviate the burden on the data movement from one memory bank to another.

In order to alleviate these challenges, Hsieh et al. [7] propose a new programmer-transparent data mapping mechanism. It co-locates offloaded code and data in the same PIM computation unit by exploiting predictability in the memory access patterns out of offloaded code blocks. Figure 7.8 shows the memory access patterns for various memory-intensive workloads selected for offloading candidate code blocks. They are backward propagation (BP), BFS graph traversal (BF), K-means (KM), and CFD solver (CFD) from Rodinia 3.0 [8], LIBOR Monte Carlo (LIB) and RAY tracing (RAY) from GPGPU-Sim [9], and Fast Walsh-Hadamard transform

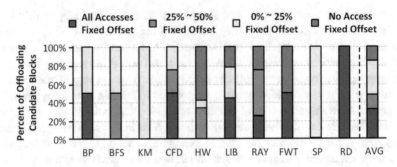

Fig. 7.8 Analysis of memory access pattern

(FWT), scalar product (SP), and parallel reduction (RD) from CUDA SDK. The result shows that 85% of all offloaded code blocks have a fixed offset between access addresses, generating a predictable access pattern. With this predictability given, observing only a small fraction, 0.1%, of initial offloading candidate instances can achieve the same effect as observing whole offloading candidate instances in data access. Although it can modify the data mapping for PIM offloaded code blocks, it keeps the original memory mapping for the rest of the data to be executed on the main CPU/GPU with maximized bandwidth utilization.

7.4 PIM Execution Scheduling

In this section, we will discuss the third challenge, the dynamic scheduling of PIM offloaded code. Section 7.2 discussed PIM offloading execution in a static manner such that a compiler statically identifies what code to offload to PIM with analytical energy and memory models. Along with the static decision, the dynamic decision can improve the optimization of the offloading code to maximize the utilization of PIM and the host processor. Especially, there are research on scheduling GPU-PIM architecture, as shown in Fig. 7.9. GPU-PIM architecture consists of multiple 3d-stacked memories and the main GPU with multiple streaming multiprocessors (SMs). The 3d-stacked memory is a PIM module; it has computation capability with SMs on its logic layer, which are the PIM computation units, and is topped with memory layers.

Hsieh et al. introduce two issues in scheduling GPU-PIM execution and propose a dynamic decision mechanism in scheduling GPU-PIM architecture. First, the authors introduce that when a large number of offloading transactions are queued for the PIM computation unit, which cannot handle fast enough, it causes a performance bottleneck. In this case, GPU is waiting on the PIM computation unit to complete whole executions. Second, they introduce that discrepancy in the bandwidth savings of the off-chip data transmissions causes the memory bottleneck. It means that offloading such transactions might only burden one of receive (RX) or transmit

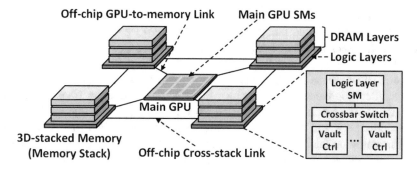

Fig. 7.9 GPU and processing-in-memory architecture

(TX) channels, while the other is left underutilized. TX is the transmit channel from the GPU to the 3d-stacked memories, while RX is the receive channel from the 3d-stacked memories to the GPU. It is necessary to look at RX and TX channels separately in deciding whether to offload blocks to PIM or not. This work proposes dynamic offloading aggressiveness control, which decides final calls on offloading codes based on runtime information. The dynamic offloading aggressiveness control adds two features in scheduling. First, it sets a limit on the number of pending offloading requests. GPU manages the pending offloading requests to each PIM computation unit and prevents further requests if it reaches their limit. Second, it sets a threshold on bandwidth utilization rate. GPU monitors the RX and TX channels not to burden off-chip data transmission. It will prevent additional offloading blocks if the utilization rate passes the threshold.

As described earlier, PIM-friendly codes can be dynamically identified rather than statically identified. The authors propose a new runtime mechanism that identifies whether the selected code blocks should be offloaded based on the conditions in GPU systems. This new mechanism selects instructions to offload to PIM computation units without any programmer's intervention. It provides seamless offloading of instructions, functions, or library calls to PIM computation units. How it works is simple with the following three steps. First, it estimates the cost-benefit of the memory bandwidth in static compile time. The expected memory bandwidth savings for different code blocks are calculated. Equation 7.1 estimates the changes in the bandwidth of TX and RX when a certain code block is offloaded to GPU.

$$\mathbf{BW}_{TX} = \mathbf{REG}_{TX} - (\mathbf{N}_{LD} + 2 \cdot \mathbf{N}_{ST})$$
$$\mathbf{BW}_{RX} = \mathbf{REG}_{RX} - (\mathbf{N}_{LD} + 1/4 \cdot \mathbf{N}_{ST}) \tag{7.1}$$

In load instruction, the address is sent through TX, and data is received through RX. In store instruction, address and data are sent through TX, and the acknowledgment messages are received through RX. **REG** represents the number of registers transferred through the RX or TX channel, and **N** represents the number of loads or stores in an executed code block. Second, it then identifies potential offloaded candidate code blocks if the result of $(\mathbf{BW}_{TX} + \mathbf{BW}_{RX})$ in Eq. 7.1 is negative. It means that the overall benefit of offloading the code block is expected to save off-chip memory bandwidth. The compiler also marks it with 2-bit tag bits indicating whether the code benefits to save RX and TX bandwidth. Third, it makes the final decision based on two runtime information. GPU keeps track of pending offloading code block requests to each PIM computation unit and monitors the bandwidth of TX and RX channels. With this information, GPU can block further requests based on their 2-bit tag bits when it exceeds each PIM computation unit's hardware limit or bandwidth utilization threshold of TX and RX channels. This dynamic final decision call for the offloading request may override the previous compiler's offloading request.

Adding PIM into a current architecture requires harmonization of kernel execution between GPU and PIM computation cores. Due to the sequential execution of the kernels, GPU and PIM suffer from under-utilization of hardware resources, even with the accurate prediction model for kernel offloading. For example, while PIM cores in memory are executing on kernels, GPU is underutilized and not running. It wastes GPU resources while GPU is waiting on PIM to finish or following GPU kernel requests to arrive. In order to solve this issue, Pattnaik et al. [10] conduct research on scheduling GPU-PIM architecture, which investigates concurrent scheduling mechanisms on multiple kernels on the main GPU and the PIM computation units in memory. It proposes two new runtime techniques: kernel offloading mechanism and concurrent kernel management mechanism, as illustrated in Fig. 7.10. The authors propose a new kernel offloading mechanism that identifies where to execute a kernel with a regression-based affinity prediction model at runtime. The regression model is built on a kernel-level analysis with three categories: memory intensity, kernel parallelism, and shared memory intensity. Table 7.1 summarizes the metrics used for predicting compute engine affinity and execution time of main GPU and GPU-PIM.

The detailed description on the three categories of predictive metrics is as follows. First, measuring the memory intensity of a particular kernel can be acquired by three metrics: memory-to-compute ratio, the number of computing instructions, and the number of memory instructions. They determine whether the

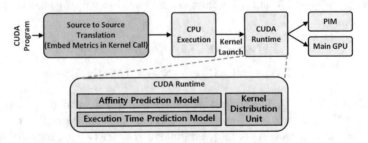

Fig. 7.10 Kernel offloading mechanism and concurrent kernel management mechanism

Table 7.1 Metrics for predicting compute engine affinity and execution time

Primary category	Predictive metric	Static/dynamic
Memory intensity of kernel	Memory-to-compute ratio	Static
	Number of compute instructions	Static
	Number of Memory Instructions	Static
Available parallelism in the kernel	Number of CTAs	Dynamic
	Total number of threads	Dynamic
	Number of thread instructions	Dynamic
Shared memory intensity of kernel	Total number of shared memory instructions	Static

kernel requires higher bandwidth or higher computing power. Depending on these metrics, the model can identify which computation cores can execute such kernel better. Second, kernel parallelism can be obtained in dynamic runtime using the number of cooperative thread arrays (CTAs). A high number of CTAs in a kernel means that the kernel has high parallelism, and a GPU with many cores can handle such kernels efficiently. Third, the shared memory intensity of a kernel is measured by the total number of shared memory instructions which tells how much a kernel reuses data. With the high total number of shared memory instructions, the kernel does not require high off-chip DRAM bandwidth. It means that the kernel has less chance of causing a memory bottleneck. In this case, the main GPU can handle better. In addition, the authors also propose a logistic regression model to predict the affinity of a kernel using these metrics, as written in Eq. 7.2.

$$\delta(t) = \frac{e^t}{e^t + 1} \tag{7.2}$$

$$\begin{cases}
\delta(t) & = \text{Model Output (0 if } \delta(t) < 0.5, 1 \text{ else if } \delta(t) \geq 0.5) \\
t & = \alpha_0 + \alpha_1 x_1 + \alpha_2 x_2 + \alpha_3 x_3 + \alpha_4 x_4 + \alpha_5 x_5 + \alpha_6 x_6 + \alpha_7 x_7 \\
\alpha_i & = \text{Coefficients of the Regression Model} \\
x_i & = \text{Predictive Metrics/Variables (Table 7.1)}
\end{cases} \tag{7.3}$$

The regression model uses a total of 25 applications, where 60% and 40% of them are used for training and testing, respectively. As a result, the model can accurately predict a kernel with 83% accuracy.

For the other runtime technique, the authors propose a new concurrent kernel management mechanism for both GPU and PIM computation units with three key information: kernel dependency information, affinity prediction model, and execution time prediction model. Kernel dependence graph is obtained by read-after-write (RAW) dependencies across the kernels by profiling the whole application's kernel execution. It helps to determine which kernels can execute in parallel. The affinity prediction model is obtained by the logistic regression model described above. It determines which computation cores can execute each kernel. The execution time prediction model predicts the execution time of a kernel on each computation core. The equation for the execution time prediction is shown in Eq. 7.4 and is obtained by the linear regression model.

$$y = \beta_0 + \beta_1 x_1 + \beta_2 x_2 + \beta_3 x_3 + \beta_4 x_4 + \beta_5 x_5 + \beta_6 x_6 + \beta_7 x_7 \qquad (7.4)$$

$$\begin{cases} y & = \text{Model Output (Predicted Execution Tim)} \\ \beta_i & = \text{Coefficients of the Regression Model} \\ x_i & = \text{Predictive Metrics/Variables (Table 7.1)} \\ & \begin{cases} 1 \text{ (very Low)} & \text{if } y < 10\text{K} \\ 2 \text{ (Low)} & \text{if } 10\text{K} < y < 500\text{K} \\ \text{Bins} \begin{cases} 3 \text{ (Medium)} & \text{if } 500\text{K} < y < 5\text{M} \\ 4 \text{ (High)} & \text{if } 5\text{M} < y < 50\text{M} \\ 5 \text{ (very High)} & \text{if } 50\text{M} < y \end{cases} \end{cases} \end{cases} \qquad (7.5)$$

The equation uses the same metrics used in the affinity prediction model in Table 7.1. The information on execution time helps balance the kernels between computation cores and minimizes the under-utilization issue. For example, if two independent kernels have an affinity toward the same computation core while the other computation core has no kernel to execute on, it suffers from under-utilization of hardware resources. In this case, it is better to offload a kernel with a lower execution time to an underutilized computation core, reducing the execution time of whole kernels and the under-utilization issue.

7.5 Cache Coherence

In a uni-processor computer system, a single core does all the work and manages the data in memories. Even with a change in a value in any memory location, it does not affect the correctness of the computation since the single core can always see latest data. On the other hand, in a multi-processor computer system, this may not always be true. When multiple processors are working simultaneously on the same data locations, they can access them freely as long as any core does not modify the data. However, once one processor modifies the data, the other processors may not acknowledge the change of the data in their local caches. This situation is when cache coherence protocol comes into play. Cache coherence protocol manages to write permission to each cache of the core to update or invalidate stale data value. It also handles the arbitration of requests from multiple cores to the same memory address. As a result, all the cores can see the valid data anytime, even if they simultaneously work on the same data.

Cache coherence mechanism could also apply in PIM architecture and is a primary challenge for enabling general-purpose PIM execution. If we consider PIM units as a conventional multi-processor architecture, we can apply traditional cache coherence protocol. PIM can be programmed as multi-core programming based on traditional shared memory with the host processor. As a result, the PIM programming model becomes simple, and PIM architecture can easily turn to general-purpose systems. However, applying traditional cache coherence protocol to PIM causes a significant overhead in off-chip memory bandwidth with many fine-grained coherence message transactions. As a result, it reverses the main benefit of PIM, high bandwidth and low latency execution. Traditional cache coherence protocol in traditional multi-processor architecture does not have this issue since it can exploit the wide bandwidth of on-chip shared interconnect. Several solutions proposed by previous researches [1, 7, 11–13] suggest some restrictions on the programming model with cache bypass policy, writeback, and message passing based mechanism. For example, Ahn et al. [11] propose to use message passing to communicate between PIM cores and CPU caches. HBM-PIM [1] and GraphPIM [13] use cache bypassing policy for offloading target. It makes a part of the memory region uncacheable and lets all memory requests bypass the cache hierarchy and send write requests directly to memory. Ahn et al. [12] use back-invalidation or writeback for the cache block before and after PIM execution. Hsieh et al. [7] propose to use write-through for cache coherence between GPU and PIM. These mechanisms could work for applications that share not so much data between PIM and the host. However, this might not always be true if tremendous data is shared between PIM and the host. All these restriction-based mechanisms could cause a degradation in performance by forcing data to write back or write through to memory frequently rather than staying in a cache.

Regarding this issue, Boroumand et al. [14] propose a new coherence mechanism called coherence for near data accelerators (CoNDA), as shown in Fig. 7.11. The authors also analyze three different existing coherence mechanisms for near data accelerator (NDA): Non-cacheable approach (NC), coarse-grained coherence (CG), and fine-grained coherence (FG).

First, the non-cacheable mechanism forces the CPU to write data to memory when the CPU has to update data, enabling NDA always to see valid data. It works well in a specific condition when the CPU hardly accesses the NDA memory, while it works poorly with most of the cases when the CPU accesses the NDA memory frequently. The authors evaluate three different graph applications from a multi-threaded graph framework called Ligra [15]: *Connected Components*, *Radii*, and *PageRank*. For the dataset, it uses *arXiv* and *Gnutella25* [16]. Figure 7.12 shows the memory system's energy consumption and speedup graph of different coherence mechanisms on different applications. As shown in the graphs, the energy consumption and the performance of the non-cacheable approach are worse than CPU-only due to frequent accesses made by the CPU threads to NDA memory. Second, coarse-grained coherence is another type of coherence mechanism that optimizes the enforcement of conventional coherence. It forces monitoring

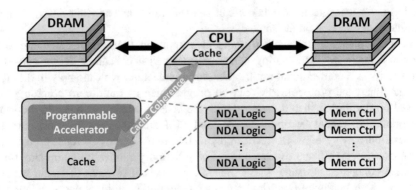

Fig. 7.11 Organization of CoNDA architecture

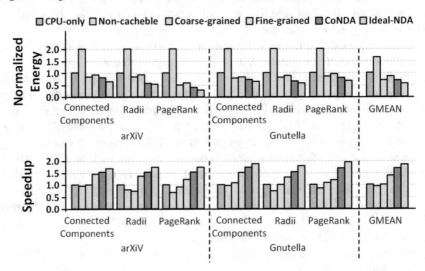

Fig. 7.12 Energy consumption and speedup of different existing cache coherence protocols

the coherence of much larger memory regions, which helps avoid unnecessary broadcasts and cache-tag look-ups. This mechanism works well on the NDA having limited shared data with the CPU threads, while it works poorly with much more shared data as it causes unnecessary data movement between the CPU and the NDA. The CPU must write all cache lines within the same NDA memory region even though the NDA only accesses a few memory addresses. As shown in the graph, coarse-grained coherence is 0.4% slower than CPU-only and is still not a good fit for NDA for many applications. Third, fine-grained coherence is a traditional protocol and works well with the applications involving irregular memory accesses. However, the limited off-chip bandwidth between the memory and the CPU cannot handle unnecessary off-chip data movements.

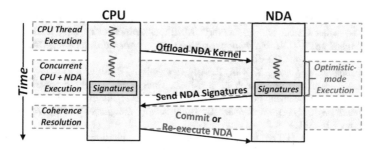

Fig. 7.13 CoNDA operation flow

Instead of applying the existing coherence protocols, the authors propose a new efficient cache coherence protocol for NDA called CoNDA, which executes NDA on optimistic execution mode, as shown in Fig. 7.13. Its new protocol works as follows. First, let the NDA always execute on optimistic execution mode. The NDA stops issuing any coherence request to the CPU during optimistic execution mode and keeps track of memory accesses. Also, it assumes that it always has coherence permission to the CPU without even looking at the CPU coherence directory. It guarantees that none of the modified data during the optimistic execution mode is written to memory. Second, after the NDA completes optimistic execution mode, it starts dealing with the coherence requests that could not be issued during the optimistic execution mode. It only works on the shared data that was actually used during the execution using the memory access tracking information, including the addresses of all NDA read, NDA write, and CPU write. CoNDA compares this information to find necessary coherence requests. Depending on the cases, CoNDA either makes the NDA invalidate or re-execute all the un-committed updates or the CPU to resolve the necessary coherence requests. The authors evaluate the CoNDA protocol by comparing it to NDA execution using NC, CG, FG, and ideal-NDA. Ideal-NDA is a reference model that does not count any coherence overhead, thus giving the best performance result. The speedup performance analysis is also shown in Fig. 7.12. It appears that both CG and NC hardly benefit from PIM due to its high cost in maintaining the coherence. FG, on the other hand, achieves 44.9% of Ideal-NDA's performance benefits. However, CoNDA achieves the most benefits among the other coherence protocols. It appears that CoNDA improves performances over CPU-only by 66.0%. The author shows that CoNDA effectively reduces the number of unnecessary coherence requests that travel through off-chip buses.

References

1. S. Lee, S.-h. Kang, J. Lee, H. Kim, E. Lee, S. Seo, H. Yoon, S. Lee, K. Lim, H. Shin, J. Kim, S. O, A. Iyer, D. Wang, K. Sohn, N.S. Kim, Hardware architecture and software stack for PIM based on commercial DRAM technology: industrial product, in *2021 ACM/IEEE 48th Annual International Symposium on Computer Architecture (ISCA)*. IEEE, Piscataway (2021), pp. 43–56

2. M. He, C. Song, I. Kim, C. Jeong, S. Kim, I. Park, M. Thottethodi, T.N. Vijaykumar, Newton: a DRAM-maker's accelerator-in-memory (AiM) architecture for machine learning, in *2020 53rd Annual IEEE/ACM International Symposium on Microarchitecture (MICRO)*. IEEE, Piscataway (2020), pp. 372–385
3. A. Boroumand, S. Ghose, Y. Kim, R. Ausavarungnirun, E. Shiu, R. Thakur, D. Kim, A. Kuusela, A. Knies, P. Ranganathan, O. Mutlu, Google workloads for consumer devices: mitigating data movement bottlenecks, in *Proceedings of the Twenty-Third International Conference on Architectural Support for Programming Languages and Operating Systems* (2018), pp. 316–331
4. Y. Kim, V. Seshadri, D. Lee, J. Liu, O. Mutlu, A case for exploiting subarray-level parallelism (SALP) in DRAM, in *2012 39th Annual International Symposium on Computer Architecture (ISCA)*. IEEE, Piscataway (2012), pp. 368–379
5. V. Seshadri, D. Lee, T. Mullins, H. Hassan, A. Boroumand, J. Kim, M.A. Kozuch, O. Mutlu, P.B. Gibbons, T.C. Mowry, Ambit: in-memory accelerator for bulk bitwise operations using commodity DRAM technology, in *2017 50th Annual IEEE/ACM International Symposium on Microarchitecture (MICRO)*. IEEE, Piscataway (2017), pp. 273–287
6. V. Seshadri, Y. Kim, C. Fallin, D. Lee, R. Ausavarungnirun, G. Pekhimenko, Y. Luo, O. Mutlu, P.B. Gibbons, M.A. Kozuch, T.C. Mowry, RowClone: Fast and energy-efficient in-DRAM bulk data copy and initialization, in *Proceedings of the 46th Annual IEEE/ACM International Symposium on Microarchitecture* (2013), pp. 185–197
7. K. Hsieh, E. Ebrahimi, G. Kim, N. Chatterjee, M. O'Connor, N. Vijaykumar, ..., S.W. Keckler, Transparent offloading and mapping (TOM) enabling programmer-transparent near-data processing in GPU systems. ACM SIGARCH Comput. Archit. News **44**(3), 204–216 (2016)
8. S. Che, M. Boyer, J. Meng, D. Tarjan, J.W. Sheaffer, S.H. Lee, K. Skadron, Rodinia: a benchmark suite for heterogeneous computing, in *2009 IEEE International Symposium on Workload Characterization (IISWC)*. IEEE, Piscataway (2009), pp. 44–54
9. A. Bakhoda, G.L. Yuan, W.W. Fung, H. Wong, T.M. Aamodt, Analyzing CUDA workloads using a detailed GPU simulator, in *2009 IEEE International Symposium on Performance Analysis of Systems and Software*. IEEE, Piscataway (2009), pp. 163–174
10. A. Pattnaik, X. Tang, A. Jog, O. Kayiran, A.K. Mishra, M.T. Kandemir, O. Mutlu, C.R. Das, Scheduling techniques for GPU architectures with processing-in-memory capabilities, in *Proceedings of the 2016 International Conference on Parallel Architectures and Compilation* (2016), pp. 31–44
11. J. Ahn, S. Hong, S. Yoo, O. Mutlu, K. Choi, A scalable processing-in-memory accelerator for parallel graph processing, in *Proceedings of the 42nd Annual International Symposium on Computer Architecture* (2015), pp. 105–117
12. J. Ahn, S. Yoo, O. Mutlu, K. Choi, PIM-enabled instructions: a low-overhead, locality-aware processing-in-memory architecture, in *2015 ACM/IEEE 42nd Annual International Symposium on Computer Architecture (ISCA)*. IEEE, Piscataway (2015), pp. 336–348
13. L. Nai, R. Hadidi, J. Sim, H. Kim, P. Kumar, H. Kim, GraphPIM: enabling instruction-level PIM offloading in graph computing frameworks, in *2017 IEEE International Symposium on High Performance Computer Architecture (HPCA)*. IEEE, Piscataway (2017), pp. 457–468
14. A. Boroumand, S. Ghose, M. Patel, H. Hassan, B. Lucia, R. Ausavarungnirun, K. Hsieh, N. Hajinazar, K.T. Malladi, H. Zheng, O. Mutlu, CoNDA: efficient cache coherence support for near-data accelerators, in *Proceedings of the 46th International Symposium on Computer Architecture* (2019), pp. 629–642
15. J. Shun, G.E. Blelloch, Ligra: a lightweight graph processing framework for shared memory, in *Proceedings of the 18th ACM SIGPLAN Symposium on Principles and Practice of Parallel Programming* (2013), pp. 135–146
16. SNAP: Stanford Network Analysis Project. http://snap.stanford.edu/

Chapter 8
Conclusion

Joo-Young Kim, Bongjin Kim, and Tony Tae-Hyoung Kim

Modern computing systems based on von Neumann architecture, which broadly consists of the processor and memory device, suffer from the data movement problem between the two devices. It becomes the major performance bottleneck of the system called von Neumann bottleneck, as the performance gap between the two separate devices gets widened. Although they leverage data locality and memory hierarchy to mitigate the bottleneck, a large fraction of time and energy is spent on just moving data from the memory to the processor for actual computations. Since the deep learning revolution of the 2010s, the world has quickly moved toward using artificial intelligence (AI) and machine learning (ML) technologies. As these new technologies involve a few orders of magnitude more data than traditional workloads, the data movement problem becomes a real challenge in modern computing systems. In addition, Moore's law that fuels the computer chip performance improvement with the process scaling is also approaching to an end.

Recently, processing-in-memory (PIM) architecture that combines the processing units into the memory has gotten attention to overcome this crisis. This unified device approach can replace expensive external data movement with much faster and cheaper internal data movement. It is a paradigm shift from processor-centric design to memory-centric design. Trying to solve the von Neumann bottleneck,

J.-Y. Kim (✉)
School of Electrical Engineering (E3-2), KAIST, Daejeon, South Korea
e-mail: jooyoung1203@kaist.ac.kr

B. Kim
University of California Santa Barbara (UCSB), Santa Barbara, CA, USA
e-mail: bongjin@ucsb.edu

T. T.-H. Kim
School of Electrical and Electronic Engineering, Nanyang Technological University, Singapore, Singapore
e-mail: thkim@ntu.edu.sg

PIM is especially effective for data-intensive workloads like AI/ML and big data applications. However, there are also many challenges ahead; the fabrication process for memory is not accessible, and chip designers need to carefully design the merged logic with the physical design constraints to maximize internal memory bandwidth. In this book, we triage and investigate the PIM technology according to the memory type it is based on: SRAM-based PIM, DRAM-based PIM, and ReRAM-based PIM. For each PIM category, we thoroughly cover the basic operation of the memory, circuit component designs, macro designs, and entire architectures and operations. The summary for each category is as follows.

SRAMs are typically used as cache memory placed near the processor in conventional digital systems following von Neumann architecture. Compatibility with the standard logic CMOS process and the low-density nature of the larger bitcell size (compared to other memory technologies) has driven the technology to be used as a small capacity and high-speed on-chip memory. SRAM-based PIM architecture naturally became a popular choice based on the solid integration compatibility that enables the additional computing feature to the regular memory operation without raising new concerns in manufacturing feasibility and the efficacy of the assigned role as a processing element. SRAM-based PIM has primary design challenges similar to the challenges in the design of SRAM for cache memory. For example, low storage density, limited bitline dynamic range, noise margin issue, and variation-induced nonlinearity are the challenges for both SRAM-based cache and PIM macros. Other challenges raised by the additional computing task assignment for SRAM PIM macros are data conversion overhead, voltage–frequency scaling, macro scalability for different DNN networks, and DNN parameter data compression. For most PIM implementation of DNN applications, the function of the PIM macro is simple multiply-and-accumulate (MAC) or multiply-and-average (MAV). However, additional mathematical operations must be processed in the preceding or following digital blocks in the system (or system-on-chip). As a result, PIM macros are rarely designed as a stand-alone processor but instead are built as a co-processor or an accelerator in larger ASICs, FPGAs, or SoCs. For this reason, input and output data of PIM macros that operate in the mixed-signal domain must provide some form of data conversion layer in its functional data flow pipeline. The data conversion in mixed-signal SRAM-based PIM is a significant design overhead in latency, energy consumption, and hardware footprint. Besides, application flexibility is another concern as the conversion blocks such as DAC/ADC are implemented in fixed precision. Custom-designed SRAM cells typically target to resolve SRAM-specific issues and utilize highly optimized supply voltage and timing schemes while trading off the configuration flexibility of the PIM macro. Data flow and architecture-level improvements are required to achieve the scalability of the PIM macro. Digital implementation of the PIM macro alleviates many of the stated concerns, but it is not a complete solution and remains an active research area. The application parameter data storage is another critical issue for SRAM-based PIM due to its low-density memory bitcell array. While there is no physical solution to map millions of high-precision parameters into the SRAM without sacrificing efficiency, algorithmic improvements can resolve the issue through data compression. Some promising approaches in this area are parameter quantization, sparse matrix operation, and multi-dimensional tensor reduction.

Using only a single transistor and a capacitor for the memory cell, DRAM has two good properties for the memory: a high capacity and a high speed. For this reason, DRAM has been used as a main memory of the computing systems and is commercially successful in the memory market. Although DRAM technology has been developed focusing on cell density, it provides a mature process solution including logic circuit implementation needed for PIM. Many approaches have been proposed to apply in-memory processing to DRAM. Based on the level of logic integration, we triaged the DRAM-based PIM into three groups: bulk bitwise PIM, bank-level PIM, and 3-d PIM. The bulk bitwise PIM integrates gate-level logic at the bitline sense amplifiers to perform row-wise processing, utilizing the maximum internal data bandwidth. However, due to the extremely narrow pitch of a DRAM cell, adding a simple logic gate, there may not be possible. Another hurdle is that it is hard to perform complex functions such as reduction, as it applies the same low-level operations on the entire row. The next option for PIM is bank-level, which integrates processing logic after column decoders in memory subarray. Since the processing logic can use the whole width of the memory subarray, not a single cell pitch, it is affordable to add more logic functions in the space. In addition, as every DRAM includes column decoders in the memory subarray, this method does not have to modify any design up to the column decoder. Understandably, bank-level PIM's maximum bandwidth out of a memory bank is the same as the regular DRAMs. Its achievable internal bandwidth is much smaller than the bulk bitwise PIM, which integrates logic before the column decoders. To compensate for this loss, bank-level PIM activates multiple banks at the same time. It also supports custom and complex commands to enable PIM operations with multi-bank activation. The types of logic added to bank-level PIM can be varied: some added dedicated logic such as matrix multiplication and the other added programmable cores. The last DRAM-based PIM architecture is 3-d PIM, which utilizes both the base logic die and memory dies in a 3-d stacked memory. It can have more design choices using multiple dies. For example, the control components in the logic die can control the execution components in DRAM dies in a 3-d vault architecture. However, the realization of 3-d PIM can be difficult due to tight physical and timing constraints among 3-d stacked dies. All the proposed 3-d PIM architectures are evaluated only using simulation. Major DRAM vendors such as SK Hynix and Samsung started to build their own DRAM-based PIM solutions. Both of them chose the bank-level PIM approach to reuse the memory subarray and minimize the changes in chip design. They also try to minimize the changes in DRAM command protocol and software as well. The performance of the first realized PIM chip on HBM is impressive but not astonishing; this is understandable considering it integrates logic at the bank level. To increase the performance dramatically, we need to integrate logic into the cell and sense amplifier level, which causes major design changes in memory subarray. We also need to discover more memory-bound applications that can benefit from DRAM-based PIM.

ReRAM-based PIMs have attracted increasing attention primarily because of its nonvolatility. Various edge computing devices will be benefited by ReRAM-based PIMs particularly when high-density ReRAM's high density and ultra-low-power consumption during the standby mode are highly demanded. However, the deployment of ReRAM PIMs in edge computing still needs to overcome several challenges. One of the most critical challenges in ReRAM-based PIM is the ReRAM fabrication technology readiness. Even though numerous ReRAM technology-related research outcomes have been reported, it is generally true that ReRAM technologies are not mature enough to be employed for commercial ReRAM-based PIM design. It is also well known that many ReRAM devices show noticeably different device characteristics such as resistance values of the high and low states, set and reset voltage levels, and disturb-free read voltage. Since the variations in the device characteristics are much more significant than that of CMOS devices, it is challenging to utilize ReRAM technology for PIMs. Another critical obstacle to be tackled is to realize the set and reset voltage compatible with CMOS technology's supply voltage. Currently, many ReRAM devices use the set and reset voltage higher than the supply voltage of the mainstream CMOS technology. Even though boosted voltage can be employed for the set and reset operations, it will create device reliability issues. The high set and reset voltages become a more critical issue when the weights stored in ReRAM-based PIMs need to be loaded or updated frequently. In contrast to the set and reset voltage, it is necessary to increase the disturb-free voltage for improving the MAC precision. In general, ReRAM-based PIMs limit the bitline voltage similar to or less than 0.1–0.3 V. This will also limit the maximum bitline current that can be generated for MAC results. Higher bitline voltage will allow more room in generating more accurate MAC results. ReRAM device endurance also affects the accuracy of ReRAM-based PIM. Even though neural networks have a certain degree of tolerance to ReRAM-based PIM accuracy degradation over time, significant endurance degradation will lead to unacceptable performance deterioration. Even though ReRAM technology issues are more fundamental, there are also other ReRAM-based PIM design challenges. In general, ReRAM-based PIMs have been used for relatively lower output precision. To overcome this limitation, CMOS design techniques for improving output precision should be more advanced. Digital ReRAM-based PIM can be a solution for improving the output precision by using almost-digital signals in the ReRAM-based PIM. In addition, it is also necessary to train ReRAM-based neural networks including the non-idealities of the ReRAM-based PIM. This will reduce the error between the trained weights using the ReRAM-based PIM and those using ideal training algorithms, which will improve the inference accuracy.

In addition, we have added a dedicated section for the PIM designs for ML training. Since the training process produces more intermediate data and requires more data movement than the inference, we think PIM has more opportunity to get performance and energy gain. Although the training process's complex computation and data flow make it hard to design, there is more room for research. Finally, we discuss the issues on the software stack to integrate the PIM hardware into a computing system. To make the PIM hardware be widely adopted in the future, its full potential should be easily exploited by end-users. For that, a renewed software stack for PIM covering programming language, library, runtime, and the device driver is essential. We hope this book can help readers understand PIM technology with a holistic view, from lower-level circuit implementation to system integration. We also hope that this book can give readers ideas and directions for their future research.

Correction to: Processing-in-Memory for AI

Joo-Young Kim, Bongjin Kim, and Tony Tae-Hyoung Kim

Correction to:
J.-Y. Kim et al. (eds.), *Processing-in-Memory for AI*,
https://doi.org/10.1007/978-3-030-98781-7

This chapter's author name was inadvertently published with a typo error. The Author Donghyuck Kim name is now updated to Donghyuk Kim. The book has been updated with the change.

The updated version of these chapters can be found at
https://doi.org/10.1007/978-3-030-98781-7_4
https://doi.org/10.1007/978-3-030-98781-7_7

Printed in the United States
by Baker & Taylor Publisher Services